The Happy Hen Handbook

The Happy Hen Handbook

A Beginner's Guide to Raising Backyard Chickens: Choosing Breeds, Building Coops, Feeding, Health Care, and Sustainable Flock Management

Written by Sophie McKay

Copyright © 2025 Sophie McKay

Published in the United States of America, 2025

MEDICAL DISCLAIMER

This book provides information about raising and caring for chickens for educational purposes only. It is not intended as veterinary, medical, or professional livestock management advice. If your chickens show signs of illness, distress, injury, or abnormal behavior, always consult a qualified veterinarian or poultry health professional.

The methods and practices described in this book—including feeding guidelines, coop maintenance, health tips, and humane processing—are general recommendations meant to support responsible chicken keeping. They are not replacements for professional veterinary care or expert intervention when needed. Never delay or avoid seeking proper veterinary assistance based on information presented in this book.

LEGAL NOTICE:

This book is copyright protected and intended for personal use only. All rights reserved. No portion of this book may be reproduced, stored in a retrieval system, or transmitted in any form or by any means—electronic, mechanical, photocopying, recording, or otherwise—except for brief quotations in a review, without prior written permission from the author or publisher. For permissions or inquiries, please contact Sophie.McKay.Author@gmail.com

First edition, 2025

ISBN 978-1-916662-51-3 (ebook)
ISBN 978-1-916662-52-0 (hardback)
ISBN 978-1-916662-53-7 (paperback)

Website: www.sophiemckay.com
Publisher: www.smartmindpublishing.com

TABLE OF CONTENTS

INTRODUCTION 11

CHAPTER 1 BACKYARD, CHICKENS, AND YOU **15**

    What Came First? Raising Chickens, Of Course! 16
    The Good, the Bad, the Clucky 17
    Myth vs Reality: What Are the Risks of Keeping Chickens? 21
    Important Points to Consider 23
    Chicken Farming Starter Pack 24
    The Cost of Raising Backyard Chickens 24
    The Takeaway 27

CHAPTER 2 BIRDS OF A FEATHER: LEARNING DIFFERENT CHICKEN BREEDS **29**

    Understanding Chicken Anatomy 30
    Choosing Your Chicken Breed 32
    Common Chicken Breeds 34
    The Takeaway 44

CHAPTER 3 A HOUSE FOR YOUR HENS **45**

    House Hunting 101 46
    Types of Chicken Coops 50
    DIY Chicken Coop 51
    The Cost of Building A House for Your Hens 56
    The Takeaway 59

CHAPTER 4 COOP CARE: CHICKEN AND EGG **61**

    Meeting Nutritional Needs 62
    Complete Feed vs. Scratch and Grain Feeding 63

Feeding Chickens by Age and Type 64
Feeding Behavior and Pecking Order 65
Food Quality Control: Maintaining Food and Water Hygiene for Healthy Chickens 67
Eggs in Your Basket 68
The Secret Life of Farm-Fresh Eggs 70
When Chickens Turn Snackers (And How to Stop It) 71
The Joy of Happy Hens: Fun Enrichment Ideas 72
Egg Abnormalities Guide: Weird Eggs 101 73
When to Worry 80
Seasonal Chicken Care 81
Winter Egg Production 83
When Winter Eggs Are a Problem 88
Winter Egg Production Quick Guide 88
All-Weather Chickens: Adapting to Your Climate .. 89
Homemade Chicken Feed 90
DIY Egg Storage Solution 93
The Takeaway 94

CHAPTER 5 HEALTHY FEATHERS, NATURALLY 95

Wing-Spotting: Telling the Healthy from the Hurting 95
Know What a Healthy Chicken Looks Like 96
Sick Chicken Treatment Options & Prevention 97
Signs of Parasites (And What to Do About Them).. 98
Prevention: The Real Secret Weapon 100
Common Chicken Diseases and Their Treatment. 103
Molting 108
The Ultimate Chicken Growers First Aid Kit 112
Monthly Chicken Health Check-Up Checklist 114
Trouble in the Coop? Quick Fix Guide 116
The Takeaway 117

CHAPTER 6 RAISING CHICKS FROM HATCH TO HOME .. 119

Preparing Your Flock For the Future 123

From Egg to Chick: Hatching at Home 124
The Takeaway 127

CHAPTER 7 THE PECKING ORDER **129**

What Is the Pecking Order? 130
Integrating New Birds – A Step by Step Survival Guide 130
Integration Checklist 139
Happy Flock, Happy Life: Reading Chicken Behavior 139
Brooders and Roosters 140
Why Roosters Rumble? 143
Pros and Cons of Keeping Roosters 144

CHAPTER 8. THE HARDEST PART: END-OF-LIFE DECISIONS **147**

After Euthanasia: Saying Goodbye 152
Raising Chickens for Meat: A Humane Approach . 154
The Takeaway 155

CHAPTER 9. CO-OP WITH YOUR COOP **157**

Catch'em And CAP'M 158
Beyond the Garden Beds: Expanding Your Flock's Natural Abilities 159
Keeping Records: Why It Matters 167
How to Keep Records (Without Losing Your Mind) 172
Sample Record Templates 174
The Takeaway 175

CONCLUSION A NOTE FROM ME TO YOU **177**

THANKS FOR READING, PLEASE LEAVE A REVIEW! **179**

BONUS CHAPTER 1 - EGG RECIPES **180**

BONUS CHAPTER 2 – DIY COOP PLANS 185

APPENDIX A 195

TROUBLESHOOTING COMMON PROBLEMS / QUICK REFERENCE 195

    Quick Fixes for Common Flock Problems 195
    Problem 10: Chickens Won't Use Nesting Boxes .. 201
    Quick Troubleshooting Flowchart 206
    The Golden Rules of Troubleshooting 207

APPENDIX B 208

HUMANE EUTHANASIA METHODS 208

    General Principles of Humane Euthanasia 208

GRAB YOUR FREE GIFTS! 217

UNLOCK THE SECRETS TO THRIVING FRUIT TREE GARDENS 218

WELCOME TO PERMACULTURE 220

BIBLIOGRAPHY 221

Looking for a gardening companion?
This Garden Planner is the perfect choice.

This guide is your trusty companion for planning, tracking, and celebrating the life in your garden, ensuring you enjoy every step of your gardening journey. Inside this logbook, you'll discover:

- **Dream Garden Planner**: Define your ideal garden and make it a reality.
- **Comprehensive Inventories**: Keep track of your **tools, seeds, roots, bulbs, shopping lists, and expenses**.
- **Seasonal Chore Planners**: Stay on top of your gardening tasks in every season, from early spring through to winter.
- **Garden Layout and Sun Map**: Plan your garden strategically.

- **Planting Timelines and Health Tracking**: Get insights on when to sow and harvest while **keeping an eye on rainfall, pests, and diseases**.
- **Pollinator Fan Page**: Celebrate the vibrant and your garden's ecosystem.
- **Tips&Tricks:** Dive deeper into gardening with **Square-Foot Gardening basics, a Companion Planting guide, the grow-bag cheat sheet** and more.

Just scan this QR code with your phone or visit the https://Gardenplanner.SophieMckay.com link to land directly on the book's Amazon page.

Introduction

I'll never forget the day when my grandparents' chickens chased me all over their farm, pecking at my legs. I was seven when I incurred the wrath of their roosters, Nugget and Reggie, by wandering too close to the barn. They fluffed their feathers and flapped their wings, eyeing me with malice. Their venomous stare and sharp beaks raised the hair on the back of my neck, and I bolted. Fortunately, my grandpa saw the chase and came to my rescue, but I was traumatized. The incident made me forever wary of winged creatures with sharp beaks.

Years later, I'm still astonished that the same scared little girl now owns hens, and even roosters. Tucked behind my home in a six-by-eight cedarwood coop is my happy flock of hens and two sharp-eyed roosters. Their enclosure is surrounded by multiple fruit trees, lying beyond rows of raised beds lush with vegetables.

My small urban farm fills my heart with joy, strengthening my connection with nature and my ancestral roots. What started off as a fun family project during the pandemic became a hobby I couldn't give up. Over the past five years, the chickens in our backyard have become an integral part of our family. Each has a name and a unique personality that sets it apart from the rest. My kids love chasing after them and grabbing them for a little snuggle. We're awakened each morning by the shrill cry of our rooster, signalling the start of a new day and I wouldn't want it any other way.

The feisty chickens in my backyard started out as golden little chicks I mail-ordered from a hatchery online. Initially, we kept the chicks in a makeshift incubator in the kitchen. We essentially fitted a large cardboard box with a heat lamp, added bedding, chick feed, and water. Once they were large enough, we shifted them to the coop in our backyard. I can see them from my kitchen window, and they never fail to amuse me with their antics. They're curious little birds, scratching and pecking at the ground in search of food. They keep me entertained all day.

The eggs they lay end up at our breakfast table, poached, fried, scrambled, or boiled, and their rich yolks give my baking a flavor that store-bought eggs can't match. As for the shells, I grind them up into a fine powder that I sprinkle onto the soil in my garden to enrich it with calcium.

I don't have to worry about egg prices going up or shortages at the supermarket. The chickens lay enough eggs to meet my family's needs, with a few extra to sell or share with neighbors. Looking back on the past five years, I can honestly say that raising chickens in my backyard has been one of the most rewarding experiences of my life.

If you've ever been frustrated by the rising cost of eggs or the empty shelves during bird flu outbreaks, you'll understand why more and more people are turning to backyard chicken keeping. In the pages ahead, I'll share what I've learned: simple, practical tips to keep your hens happy and healthy so that you can enjoy a steady supply of fresh eggs right from your own yard.

In Chapter One, we'll look at the advantages and disadvantages of raising hens, along with the estimated costs of getting started. Chapter Two introduces the different chicken breeds and helps you choose the right ones for your needs. Chapter Three is all about finding a suitable home for your flock. In Chapter Four, we'll explore how to care for both your chickens and their eggs, while Chapter Five focuses on keeping their feathers healthy and clean. Chapter Six covers everything you need to know about raising chicks, and Chapter Seven helps you decode chicken behavior and understand the pecking order.

Finally, Chapter Eight focuses on building sustainable systems to care for your flock, such as composting manure and using your chickens to control garden pests. And as a bonus, I've included an extra chapter at the end with my favorite egg recipes and DIY coop plans.

Our flock has grown significantly over the past few years. We started with just three chicks named Dorothy, Anabelle, and Nancy. Five years later, they've turned into gorgeous, plump hens that love strutting around the backyard. Our initial success gave us the confidence to add more to the flock. Of all the chickens, Dorothy is my favorite. She's one noisy hen, full of sass, and loves to follow me around the yard.

Suffice it to say that I adore my chickens and their quirky personalities as much as I love the eggs they lay. I'm a firm believer that once you try fresh eggs laid by homegrown chickens, there's no going back. The yolks are rich and bright orange, and the flavor is unmatched. There's nothing quite like the excitement of waking up in the morning and checking the coop to see how many eggs are waiting for me.

Raising backyard chickens is easier than you might think. This book is your complete guide to getting started with these adorable and surprisingly useful pets right at home. So, let's get started and crack open the secrets to happy hens and fresh eggs!

Chapter 1

Backyard, Chickens, and You

Sunny-side up, scrambled, poached, or hard-boiled, eggs are a breakfast staple in my house. They're packed with nutrients, vitamins, and good fats, perfect for a protein-rich start to the day. While I can't imagine a hearty breakfast spread without them, the reality is that the soaring price of these protein powerhouses is turning them into a luxury few can afford.

In recent years, the price of eggs has skyrocketed, jumping from US$1.20 for a dozen eggs in 2019 to US$4.15 in 2025[1]. While the escalating cost is deeply concerning on its own, it's not the only thing consumers have to worry about. Not only are eggs becoming more expensive, but they're also becoming harder to find at the supermarket due to a supply shortage. Even some restaurants have started imposing additional charges on egg-based dishes.

A significant factor driving prices up and causing a shortfall in production is avian influenza, also known as bird flu. Starting in February 2020, the outbreak has resulted in over 166 million birds being slaughtered in the U.S., accounting for nearly 98% of chickens produced on factory farms.

The rising cost of eggs, frequent shortages, and the COVID-19 pandemic have pushed more and more Americans toward raising their own chickens. According to the American Pet Products Association, a whopping 12 million people in the U.S.[3] own backyard chickens. While the idea of owning egg-laying hens is becoming more appealing for Americans, it remains a daunting venture, especially for people lacking the time, space, and experience required to rear chickens.

Unlike most farm animals, the appeal of raising chickens is that they're a fairly easy gamble. They don't take up a lot of space and are pretty low-maintenance, requiring basic food and shelter. For all the effort you put in, you're rewarded with fresh eggs that far surpass even the best organic varieties at the grocery store.

What Came First? Raising Chickens, Of Course!

So what makes raising chickens in your backyard worthwhile? The benefits include more than a consistent supply of fresh eggs that are rich in color and flavor. If you're a gardener, you can trust your feathered friends to protect your garden by devouring the pests and insects harming your plants.

But that's not all that chickens love to feast on. Got bits of vegetable peels leftover from making dinner? Just toss them inside the chicken coop and watch your feathered friends gobble them up. And when it's time to give the coop a deep clean, gather the chicken manure and sprinkle some on your plants. It's packed with nutrients that plants love, making it a natural fertilizer and saving you a bit of cash and a trip to the gardening center.

The upside of raising hens doesn't end here. The economic benefits of keeping chickens far outweigh the nutritional advantages. In 2016, Bill Gates hailed the poverty-fighting power of poultry on his website. He wrote: *"It's pretty clear to me that just about anyone who's living in extreme poverty is better off if they have chickens."*[4]

According to him, the best decision someone could make to improve their life and lift themselves out of poverty was to raise chickens. This applied to people making as little as $2 a day. His rationale for the advice included the relative ease of taking care of the birds, the low cost of maintenance, and enormous returns. Most breeds eat whatever they find on the ground and need fewer vaccines. They're fairly inexpensive to look after and grow rapidly in number. Someone starting with five hens and a rooster can end up with a flock of 40 chicks in less than three months. Selling the hens for $25 each can bring in a profit of $1000 or more annually.[4]

So why raise chickens? Because they're inexpensive, make great pets, provide you with breakfast, and bring in profits. These quirky, lovable, and beautiful animals will add life and vigour to your home. They're easy to care for, requiring minimal time and effort, and they have their own individual personalities.

The Good, the Bad, the Clucky

Although raising chickens dates back thousands of years, it wasn't until the early 20th century that the practice was widely adopted in urban areas in America. During World War I, the U.S. government encouraged families to raise poultry to bolster food supplies for the troops and allies. Backyard flocks resurged in popularity once again during the Great Depression and World War II, helping families through food shortages and rationing.[5]

After the war, interest in raising poultry declined as supermarkets expanded and industrial farming made eggs cheap and widely available. In recent years, concerns over sustainability, climate change, and food quality have sparked a renewed interest in backyard chickens. The COVID-19 pandemic accelerated this trend, as supply chain disruptions, empty store shelves, and rising egg prices drove more people to raise their own hens.

Five years ago, I decided to keep chickens in my backyard. While I consider it the best decision I ever made and adore my feathered friends, I advise others to weigh the pros and cons before starting their own flock. So let's look at the upsides and downsides of raising hens.

Pros

- *Control Pests:* Tired of slugs, snails, and insect larvae wreaking havoc on your plants? You can count on your backyard chickens to put an end to the pesky pests. They'll gobble them up cheerfully, protecting your plants. Insects are also an excellent source of nutrition for chickens, improving the quality of their eggs.

- *Reduce Food Waste:* You can make a hearty meal for your hens with your kitchen scraps. Fruits and vegetable peels, some pips and seeds, leftover pasta or rice are some items they love best. While it won't meet all of their nutritional requirements, it certainly adds diversity to their diet while reducing food waste in your home.

- *Get Rewarded with Fresh Eggs:* You'll have an uninterrupted supply of organic eggs rich in flavor and quality, far superior to anything you might find at your local supermarket.

- *Get Rid of Weeds:* Let your hens loose in the garden and watch the magic happen. As they peck, scratch, and dig, they naturally uproot weeds and gently till the soil, doing the hard work for you.

- *Easy to Care for and Maintain:* They're surprisingly easy to care for and maintain. All you need is a bit of self-discipline to keep your flock well-nourished and healthy while keeping their living space clean and secure.

- *Produce and Excellent Fertilizer:* Chicken manure makes an excellent fertilizer. Simply add your hen-house bedding to your compost bin to create a nutrient-rich fertilizer.

- *Make Great Pets:* These adorable birds have unique, quirky personalities, making them excellent pets. Children love them. They're fun and entertaining to watch. Whether you're observing their silly antics or chasing them around, you'll be thoroughly entertained.

Cons

- *Require Space:* Although they don't take up a lot of space, you'll need at least 4 square feet per bird. Overcrowding causes the birds to become stressed, leading to increased pecking and the spread of sickness.
- *Make a Lot of Noise:* You might want to make sure you won't run into trouble with your neighbors because they can be rather noisy.
- *Require Moderate Expense:* Raising chickens in your backyard requires a few upfront costs. You'll need to invest in a good-quality chicken coop, bedding, and feed to get your chickens off to a good start. The expenses pay off once your hens start laying eggs, cutting down your grocery bill.
- *Produce Chicken Odor:* You'll have to get used to the characteristic chicken odor. The smell persists despite cleaning the coop regularly, which you may find unpleasant at first.
- *Attract Predators:* Be prepared to ward off foxes, raccoons, weasels, and badgers that hunt chickens and their eggs. You'll have to be vigilant and protect your flock as best as you can, especially during the night.
- *Require Proper Manure Management:* Chicken excrement must be disposed of properly, or it can be a health risk. It's essential to have a secure location to dispose of it.

Raising hens in your backyard can be an immensely rewarding experience, if done right. It's important to carefully consider the pros and cons before you set out to purchase your chickens. Here are five questions you must ask yourself before farming hens. **How many chickens do you plan to get?**

Chickens are highly sociable birds that require company. It's best to start out with a pair or more. Two to three chickens can lay anywhere between 10 to 20 eggs per week, a sufficient amount to fulfill the breakfast needs of a small family. Four hens or more can lay anywhere between 16 to 28 eggs each week, enough to satisfy the food requirements of a large family. Double the number of chickens, if you want enough eggs to sell them by the dozen.

1. **How much space do you have available?**

While hens don't take up a lot of room, they need at least a coop that's big enough to provide 4 square feet per chicken. A smaller coop may work if you plan to let your chickens wander freely outside.

2. **Do you find them adorable?**

It'll make your work a whole lot easier if you do. Chickens are adorable creatures that amuse their owners with their antics. If you plan to keep them in your backyard, but don't care for the birds much, it'll be tough for both you and them. And if you adore your feathered friends but your partner doesn't, then you may have to take on the bulk of the duties to care for them.

3. **Do you have the time available to take care of them?**

While they're notoriously low maintenance, they do require some daily care and regular upkeep. You'll need to put aside an hour or two to look after them.

4. **Is it legal to keep chickens in your town?**

Make sure to check your local authorities. While most U.S. towns have no restrictions on keeping chickens in your backyard, some cities impose certain limitations on the practice. Researching the local noise laws will also help you avoid running into trouble with your neighbors.

While the upsides far exceed the downsides, it's important to assess the difficulties and plan accordingly before you commit to raising chickens in your backyard. It's important to note that most of the disadvantages mentioned above can be mitigated by investing in a good-quality coop. Armed with the right information and a clear idea of what's required of you, you're likely to do a stellar job caring for your flock and reap the many rewards of raising hens in your backyard.

Myth vs Reality: What Are the Risks of Keeping Chickens?

It's understandable to have concerns about public health safety and community well-being when it comes to raising backyard chickens. However, it's important to separate fact from fiction, weeding out the myths and misinformation about raising poultry. Here are some common misconceptions and concerns about keeping hens on your property.

A Source of Disease

A 2011 review by researchers at the University of Pennsylvania concluded that backyard poultry played a minimal role in past outbreaks of bird flu. They used several case studies to back up their opinion, including the 2004 example of the bird flu epidemic in British Columbia, Canada. During the outbreak, commercial birds were 5 times more likely to get infected with the virus than backyard flocks. Similarly, not a single backyard flock was infected in the 2002 wave of avian influenza. Based on these findings, it's safe to assume that homegrown chickens are less likely to be a cesspool of disease.

Public Health Concerns

Diseases such as salmonellosis can spread from backyard chickens to humans. For instance, if your pet dog consumes chicken fecal matter, it could transmit salmonellosis to the family members. This is usually not a problem when the flocks are well-maintained and the coop is cleaned regularly. Similarly, there have been no cases of humans contracting bird flu from chickens so far in the U.S.

Waste Disposal

Proper waste management is crucial for minimizing disease risk and unpleasant odor. Most cities limit backyard flocks to a small number of birds so the waste does not become a community hazard. It's also important to keep the chicken enclosure away from important water reservoirs to avoid environmental pollution, which usually isn't a problem in urban centers.

You can estimate the amount of manure that your flock will produce with a simple formula. The weight of fresh chicken manure is 115% of their total dry feed intake. Calculate your flocks' total feed consumption per day and multiply it by 1.15. This will give you an idea of the amount of manure you'll be dealing with.[7] You can also compost the manure-filled bedding from the coop to create an excellent fertilizer for your plants.

Attract Pests and Rodents

Your cluckers are a target for a number of pests, including lice, bedbugs, ticks, mites, and fleas. Some other insects and critters you might have to worry about include moths, cockroaches, flies, beetles, and rodents. A clean and secure chicken coop can help keep these uninvited guests at bay.

Dealing with the Noise

In my house, we don't need alarm clocks to wake us up early in the morning. Our rooster is in charge of that. Each morning, his loud crow echoes through the house, waking us up. While I'd choose the shrill cry of my rooster over the rattle of an alarm clock or the drone of my phone's ringtone any day, most city regulations do not allow keeping roosters in the backyard. Some communities consider hens a source of noise nuisance as well, due to all the squawking they do, especially while laying eggs. Minimizing the number of hens you keep can help reduce the noise.

Produce Unpleasant Odor

Chicken coops have a characteristic odor. Keeping the coop clean, proper waste disposal, and composting the bedding can help minimize foul odors.

Important Points to Consider

It's important to adhere to your local laws and regulations before you head over to your local farm supply store to buy your flock. Depending on where you live, you may need a permit to build a chicken coop on your property. For example, people living in a residential zone in Huntington, New York, require building permits to modify any aspect of their house or surrounding structure. There are usually strict rules regarding backyard chickens in urban and commercial areas. It's best to check with your local authorities on the following points before you make a purchase.

1. The maximum number of chickens you can have in the backyard. Most areas allow residents to keep anywhere between five to six hens.

2. Whether you can keep roosters or not. Although they're not allowed in most urban areas because of the noise, some city bylaws make an exception for breeding purposes.

3. Several cities require hen owners to obtain a permit. Make sure you check if you'll need one, and be aware that there may be a fee. Look up the application process to see whether you'll need your neighbors' approval as well. Some communities even require prospective hen owners to take a course on rearing chickens.

4. Some areas may impose coop restrictions, such as keeping it away from neighboring properties. It's best to check with your local authorities whether this is the case before deciding on the location for your coop.

Chicken Farming Starter Pack

It's important to do your homework before you mail-order chicks from your local hatchery, so you face fewer hiccups later on. Think about the breeds you want to grow. Chickens have more than 400 varieties. Their size and egg-laying capacity vary depending on the breed. Some are raised for their meat, others for their eggs. Climate plays a huge role as well. Mediterranean breeds are better suited to hot and humid climates, while American breeds do better in cooler climates.

The area you select for the coop will have a direct impact on the well-being of your flock. It must protect the birds from weather fluctuations, predators, and theft, so think carefully about the site you choose. It should typically be located near your house on high ground, keeping the chickens dry and protected from predators and harsh winds. It can either be movable or fixed.

Once the housing is ready and you've chosen the breeds you want to grow, it's time to bring in the baby chicks. There are numerous mail-order hatcheries online that will ship the chicks to your doorstep via the U.S. Postal Service. Some have a minimum number of chicks that you can order. Some may charge a fee for choosing the sex of the little cluckers, while others only ship straight-run chicks, in which case you get whatever hatches out of the eggs.

The Cost of Raising Backyard Chickens

Raising chickens in your backyard has low startup costs. Your biggest expense will be the coop. Let's round up all the items required and their associated costs.

1. **Coop ($100 to $400)**

Depending on the kind of coop you choose, the cost can range anywhere between $100 to $4000. Factors affecting the cost include whether you buy a new or used structure. The size of the coop can also add to the price as well. If you have the skills for it, you can build the structure yourself, which will significantly cut down the expense.

You can buy used chicken coops on Facebook Marketplace or Craigslist for a couple of hundred dollars. However, if you're in the mood to splurge, then there are numerous custom chicken coop builders out there who can build a gorgeous coop to match your backyard aesthetics. These tend to be a little pricey, ranging from $2000 to $4000.

Another factor that can add to the cost is whether you want the structure to be mobile or stationary. Mobile chicken coops are also called chicken tractors. They can feature wheels or be designed to slide on rails or skids and come with a slightly bigger price tag as compared to standard stationary coops. A standard fixed structure usually costs between $500 to $1000.

2. **Predator Proofing ($50 to $100)**

Protecting your chickens from predators is essential. You'll need an apron of hardware cloth laid along the base of the coop to avoid digging critters from getting inside. Predator-proofing the area can be pricey and time-consuming, but it pays off in the end by keeping your chickens protected.

3. **Feed and Water ($50)**

You'll need at least one average-sized feeder and waterer as you start out. They'll cost $25 each. Bigger-sized containers tend to be more expensive than smaller ones.

4. **Winter Supplies ($50)**

If you live in an area where temperatures plunge below freezing during the winter season, you'll need some additional supplies to keep your chickens warm. Containers with heated bases will make sure their drinking water doesn't freeze. You can find these at your local farm supply stores. You can insulate the coop with plastic sheets or bales of straw, keeping out frosty drafts.

5. Brooder ($20 to $40)

Starting off with baby chicks gives new chicken keepers a chance to bond with their birds. While you don't need a degree in animal husbandry to raise baby chicks, you'll need to invest in a few items to attend to their basic needs. A brooder is a small structure that provides chicks with food, water, dry bedding, and a source of heat.

If you only have a handful of chicks, you can repurpose a plastic storage box or a heavy-duty cardboard box. If you don't have a container lying around your house that's big enough to house chicks, you can buy one for $10 or less. Next, you'll need a heat lamp and bulb for $20 and a bag of bedding for $10. Your DIY brooder will be ready for just under $40.

6. Chicks ($2 to $10)

The price of chicks varies depending on where you live and the options available to you. You can buy them for as little as $2 each from your local hatchery or local farm supply store for $4 to $6. Another option is to buy pullets or young chickens 4 to 16 weeks of age that haven't started laying eggs yet. They will cost you $15 to $35, depending on their age and breed.

7. Fencing ($100 or more)

Your chickens will need a large open area to run about that is separate from the coop. Many commercially available coops feature a built-in enclosed area. You might want to include a run if your coop doesn't have one or if you plan to build the coop yourself. Runs are usually constructed with wood, metal, or PVC frame that is lined with netting, hardware cloth, or chicken wire to keep the predators out.

Alternatively, you can also free-range your chickens, depending on your location. However, you may need to fence the area to keep them from wandering too far. Installing fence posts or snow fencing can cost $100. Electro-netting or an electric fence energizer can make the costs go up.

Total Cost

If you stick to the basics, your start-up cost will come to around $720. This includes five to six baby chicks bought from your local farm supply store, a basic brooder, a standard fixed coop without a run or fencing, predator proofing, a feeder and waterer, and basic winter supplies.

The Takeaway

There are numerous benefits of raising chickens in your backyard, and the few cons can be easily mitigated. The venture carries a low start-up cost while delivering huge returns. You don't need a ton of experience to keep hens on your property, just curiosity and a willingness to learn. Now that we've covered the basics, let's shift the spotlight to the stars of the show: the birds themselves.

Cost/Benefit Analysis

Here's a simple cost/benefit tracking sheet to help you estimate the total costs of setting up a chicken coop. The last section allows you to compare your profits with your total expenses to give you a better idea about whether you're moving in the right direction.

Cost/Benefit Tracking			
Variable Costs	Cost Per Unit	Total Units Required	Total Cost
Layer feed			
Water			
Bedding			
Chicks			
Chick feed			
Fixed Costs	Cost Per Unit	Total Units Required	Total Cost
Coop			
Feeder			
Drinker			
Heat Lamp			
Fencing			
Revenue	Revenue per unit	Total number of units	Total Revenue
Eggs			
Meat			

This gives you a realistic snapshot of what it takes to get started and what you can expect in return. As you fill it out, you'll begin to see how manageable the investment really is, and how quickly your hens can pay you back in fresh eggs, rich compost, and daily enjoyment. With the numbers laid out, you're ready for the fun part: choosing the right birds and welcoming them into your backyard flock.

Chapter 2

Birds of A Feather: Learning Different Chicken Breeds

The next time you're about to crack open an egg to make an omelet, pause for a second and look at its color. Is it white, brown, or blue-green? While some people mistakenly believe that the color of the egg is an indication of its nutritional content, deeming brown eggs healthier than white eggs, the reality is quite different. Egg colors are affected by chicken breeds. Despite differences in color, all eggs are nutritionally similar.

If you're someone who hasn't been around a farm much, you might have trouble telling one chicken breed apart from another. To make things easier, we'll start things off by understanding chicken anatomy. You'll learn the purpose and function of different chicken body parts and what they're called. So let's get to know our chickens and what sets them apart from other birds.

Understanding Chicken Anatomy

A good understanding of chicken anatomy is important to keep an eye on your flock's health, so let's explore the defining features of your little clucker.

Crop

This is your chicken's storage bag. It is a muscular organ located at the bottom of the chicken's neck, and it's where chickens store everything they consume. It's usually flat and empty in the mornings and grows in size as the chicken eats throughout the day.

The crop can occasionally become blocked. If this happens, you'll need to address it right away by helping the chicken clear the obstruction. Offer a small amount of water, administer a bit of olive oil to help loosen the blockage, and gently massage the crop to encourage it to pass.

Gizzard

The gizzard is a strong, muscular part of a chicken's digestive system located just past the stomach. It plays an essential role by crushing food and helping the body absorb nutrients. As chickens peck at the ground, they naturally swallow tiny stones, sand, and small pebbles. These stay inside the gizzard and act like grinding tools, helping break down food efficiently.

Oviduct

The oviduct is a long, tube-shaped organ about 25 to 27 inches in length, located along the backbone between the ovary and the tail. It plays an important role in forming an egg. When a yolk is released from the ovary during ovulation, it enters the oviduct, where the rest of the egg: egg white, membranes, and shell, gradually develop around it. This process happens whether or not the egg is fertilized.

Cloaca

The cloaca is a multi-purpose chamber where the digestive, reproductive, and urinary tracts meet. Both excrement and eggs pass through it, but each system enters the cloaca through its own opening, which helps prevent contamination. The cloaca also plays an essential role in mating, acting as the point of contact during reproduction.

Comb

The comb is the most prominent feature on a chicken's head, and along with the wattles, it's red, soft, and warm to the touch. Because chickens don't sweat, they rely on their combs and wattles to release excess body heat. These structures are rich in blood vessels and help regulate temperature by acting as a natural cooling system during warm weather.

Wattles

They're the fleshy red flaps of skin hanging down from either side of the beak. They're more prominent in roosters than hens and help regulate body temperature by releasing heat.

Ears

Chicken ears are simple openings hidden under feathers on the sides of their heads. While they lack external ears like humans, they do have prominent earlobes. The color of the lobe varies depending on the breed of the chicken.

Eyes

The eyeballs are surrounded by skin folds called eye rings. A third eyelid, known as the *nictitating membrane,* slides over the eye horizontally, protecting it from dust and debris. Their world is a lot more colorful than ours, thanks to the four types of cones (photoreceptor cells) in their eyes. They can see green, blue, and red light, but ultraviolet light as well.

Feathers

The feathers provide a form of insulation, keeping the chickens warm during winter as well as protecting them from harsh sunlight. Like eggs, feathers are primarily composed of proteins. During molting, chickens stop laying eggs as their body redirects nutrients toward producing new feathers. Each feather arises from a follicle and follows a particular pattern on the body. For instance, some areas remain featherless, called *apterylae*. New feathers appear as pins. They're covered in a keratin sheath that eventually falls once the feathers are fully developed.

Choosing Your Chicken Breed

Raising chickens in my backyard was a big decision. I remember puzzling over which breed to buy. Back then, I had no idea that it's best to choose chicken breeds based on your needs. My first chicks were bought based on looks alone, and though I love them dearly, I handpicked the rest of my flock a lot more wisely. So let's look at the different factors you should consider when choosing your birds.

Eggs? Meat? Or Looks?

The first step is to determine what you hope to gain from keeping chickens. Is it for the eggs? Meat? Or simply because you find the winged creatures absolutely adorable. Maybe it's all of the above, in which case you need a breed that fulfills each requirement. Each breed has its star feature, whether it's laying a lot of eggs, providing meat, or both.

If you want the birds for their eggs, then you'll need laying hens. These breeds require small private areas in the coop called nest boxes to lay eggs. If your primary motive for keeping the birds is the meat, then you'll need breeds that mature fast, requiring minimal expense to rear them. These include broiler and fryer hens. These birds are usually not good at laying eggs, and some may even have difficulty reproducing.

If looks are all that matter, then bantam chickens or large fowl breeds may be the ones for you. While these birds may not be the plumpest or lay the most eggs, they're sure to catch your eye by being cute as hell.

The good news is that you don't always have to pick and choose. Dual-purpose varieties offer multiple benefits. They include breeds that not only lay plenty of eggs but also provide ample meat while being pleasing to look at.

Size Matters

The bigger the chicken, the bigger the coop. Depending on the breed, chickens can be of varying sizes from the tiny Serama to the enormous Jersey Giants. Larger birds take up more room, requiring a larger-sized coop. If you have only a few square feet of space to spare, then it's best to go for a small-sized variety such as bantams. The bite-sized birds are cute as a button and can fit easily inside an 8 by 8 coop; however, their eggs tend to be smaller than other breeds

Apart from bantams, you can also find small and medium-sized breeds that require 3 to 4 square feet per bird, but lay larger eggs. Large-sized birds produce plentiful meat and eggs but require 4 to 6 square feet per hen. Very large chickens, like the Brahma roosters, can reach up to 2 feet and 6 inches, requiring up to 8 square feet per bird.

Fair-Weather Friends

Some breeds aren't suited to certain weather conditions. Based on their origin, they may or may not be able to tolerate certain climates. For instance, plump or feather-footed varieties are cold-hardy breeds best suited to survive harsh winters, but may overheat in warmer climates. Leaner, more nimble breeds originating from the Mediterranean or tropical regions fare better in warm conditions.

Eggs Galore

How big and how many eggs you want your hens to lay daily or weekly is something you must take into consideration while selecting the breed of your chickens. Keep in mind that egg yield decreases with age. There are breeds that lay 0 to 50 eggs per year. These birds are kept mostly for show and include crosses with wild jungle fowls.

Then there are breeds capable of laying several hundred eggs each year, laying one or more eggs each day. These include Lohmann Browns that lay up to 300 eggs a year. It's important to provide these birds with layer feed, a specially formulated diet rich in calcium and protein, to support strong shells and consistent egg production.

When it comes to egg color, you'll be surprised by the incredible variety chickens of various breeds have to offer. Blue, white, green, brown, speckled, or chocolate, there's an endless list of colors available. A peek at your hen's earlobes can give you some idea of the kind of eggs you should expect. Oftentimes, the color of the chicken's earlobe matches the color of the egg.

Common Chicken Breeds

Choosing the right breed for your backyard can make caring for the flock a lot easier. Hens that are suited to their environment are much happier and healthier. Some factors to consider while selecting breeds for your coop include their purpose, available space, hardiness, health, and temperament. So let's look at some common breeds perfect for your backyard.

Rhode Island Red

A popular choice for backyard coops, these birds are known for their excellent egg production and ability to tolerate both hot and cold temperatures. These friendly chickens are great for both eggs and meat. They're coveted for their striking red plumage. The roosters can weigh up to 8.5 lbs. The hens are somewhat lighter, weighing around 6.5 lbs.

They can adapt to various climates and are particularly well-suited for cold weather due to their dense plumage. However, it's important to provide them with shade and fresh water during the summer months to prevent overheating. They're notoriously low-maintenance. Being excellent foragers, they don't require any special diets. They thrive in free-range environments and do well in confined spaces provided they have enough room.

Rhode Island Reds are known for their great egg-laying abilities. They're consistent layers, producing 250 to 300 large eggs annually. They'll continue laying during the winter, although the number of eggs they produce may drop due to shorter daylight hours. Their eggs are excellent quality, large-sized, and brown in color. Their friendly temperament and impressive egg-laying ability make them a highly sought-after breed.

Plymouth Rock

These versatile, dual-purpose chickens are prolific layers. Their eggs range from large to extra-large and are brown in color. Popular varieties include the White and Barred Rocks, named for the color of their plumage. Their barred varieties have alternating bands or stripes of colors, kind of like a zebra. Their eye-catching appearance, cold-tolerance, and great egg-laying capacity make them ideal for the backyard.

They're renowned for their curious and friendly nature. They're easy to manage and cold-hardy. The hens start laying eggs around five months of age. This makes the breed one of the earliest maturing layers. The hens go through a molting period several months after laying. During this time, they lose their feathers, take on a rather scruffy appearance, and stop laying eggs. Their wattles and combs also lose color, turning pale. The molting period typically lasts from four to eight weeks. Get the egg basket ready as soon as the wattles and comb turn bright red again and you hear them sing.

The Plymouth Rock chickens are able to adapt to a number of different climates. They do well in both hot and cold weather. Their feathers provide decent insulation against the cold; however, extra care must be taken in the winter to provide them with a warm and dry place to sleep. They're susceptible to lice, mites, and respiratory problems, like most poultry, if not properly cared for. A dry, clean, and well-ventilated living environment will keep them healthy and happy.

They're relatively low-maintenance, thriving on the basics: clean water, nutritious feed, space to roam, and a secure place to rest. These skillful foragers are well-suited for free-range environments. Their remarkable egg-laying ability makes them perfect for backyard flocks. They can produce up to 200 to 280 medium to large-sized eggs annually while requiring minimal upkeep. While egg production slows down during the winter months, it rarely stops. These friendly and docile birds get along well with humans and other chicken breeds, making them the perfect addition to your diverse flock.

Orpingtons

Known for their docile and friendly nature, these birds lay large brown eggs and are able to survive low temperatures. Buff Orpingtons, celebrated for their golden feathers, are the most popular variety. These gentle, dual-purpose hens possess a calm disposition, making them perfect for family farms, homesteads, and beginners. Their superior egg-laying capabilities and quality meat make them highly sought after.

It's best to keep 2-year-old Buff Orpingtons for their meat and 1-year-old hens for egg production. This breed of chickens is great at raising their chicks and mothering their young. They're perfect for those seeking easy-to-handle, low-maintenance, dual-purpose breeds.

Their quiet and friendly personalities make them a great choice for households with children. These hardy and resilient birds are renowned for their ability to adapt to a number of climates, especially cold weather, due to their dense plumage. They're fairly disease-resistant, requiring access to fresh water and a clean, well-ventilated living space to keep them free from mites and lice.

Leghorns

These prolific egg layers are known for their unique disposition. They're active, alert, and can be somewhat flighty. The most common variety, White Leghorns, can produce 300 or more eggs a year. The breed originated in Italy, Denmark, and England, and is known for being hardy and athletic. The hens lay large to extra-large white eggs.

They're quick to mature and consistently lay eggs all through the year. Being active and efficient foragers, they make excellent free-range chickens. While coveted for being exceptional layers, they're not the friendliest birds and are known to avoid human contact. They have rather nervous and flighty personalities, and they don't offer a lot of meat.

Ameraucana

Purebred Ameraucanas lay blue and green-colored eggs. They're friendly and come in unique varieties with tufted facial feathers called muffs and beards. Highly regarded for their egg-laying capability, these chickens have a docile and friendly nature, ideal for homesteads, family farms, and beginners. They adapt readily to coops and free-ranging environments.

The Ameraucanas have slate blue legs, medium-sized pea combs, well-spread tails, and red earlobes. They remain a favorite among poultry enthusiasts due to their ability to adapt to a wide range of climates, their inquisitive nature, and their extended laying season, which continues well into the winter months.

Easter Eggers

Originating from Chile in South America, the Easter Eggers are as visually stunning as the Ameraucanas. They lay colorful eggs, ranging from pale to dark blue, sometimes with shades of green and pink. Their eye-catching looks, high egg production, and delicious meat make them perfect for backyard flocks. They thrive in coops and free-range environments.

While these stunning birds adapt well to a number of climates, they require some extra care during extremely hot or cold weather. Their feathers are not as dense as other breeds to keep them warm during harsh winters. A warm, dry, and draft-free coop with plenty of access to fresh water can help them thrive in extremely cold areas.

Wyandottes

These cold-hardy birds come in various colors and possess zen personalities. They're known for their calm temperament and moderate egg-laying ability. The Blue Laced Red Wyandotte is a popular breed with rich red feathers laced with blue. The Black Laced Silver is a gentle bird with dazzling black-laced white feathers. Its excellent mothering abilities, medium to large eggs, and ability to thrive in various environments make it ideal for backyard flocks.

Silkies

Their fluffy, soft feathers and friendly personalities make them excellent pets. Although they're not the best egg layers, their small eggs are often considered a delicacy. The Buff Silkie Bantam is prized for its soft, fluffy feathers and gentle disposition. Their affectionate and calm nature makes them excellent family pets.

Sussex

They come in various colors and are excellent at laying eggs. The Speckled Sussex is the most popular variety. Their friendly and curious personalities make them endearing to their owners. They're surprisingly hardy and easily adapt to a number of different climates.

Their gentle nature, vibrant appearance, egg-laying ability, and delicious meat make them widely sought after by chicken owners.

The Speckled Sussex are consistent layers that perform well in confined spaces. Their plumage acts as a natural camouflage, protecting them from predators like foxes and coyotes. Interestingly, the number of speckles on their feathers increases with each annual molt, giving them an even more striking appearance as they age.

These lively and friendly chickens thrive in various environments. Their calm temperament and fondness for humans make them great pets. The hens make excellent mothers to their young ones. Their ability to adapt to cold climates and remarkable foraging skills help them thrive all year long.

New Hampshire Red

Developed in the early 1900s, these docile breeds originated from the Rhode Island Red in New Hampshire and Massachusetts. The New Hampshire Reds were created for faster feathering, maturity, and growth. They played a significant role uplifting the poultry industry after World War II, even leading to the creation of the Delaware chicken. In 2018, these calm-natured, dual-purpose chickens were declared New Hampshire's state birds.

Their vibrant red plumage is sure to catch the eye. These stunning birds are not only great at laying eggs but also provide ample meat. They lay 200 to 220 large eggs annually. The roosters can bulk up to 8.5 lbs, while the hens can reach a decent 6.5 lbs.

Their docile nature makes them beginner-friendly. With proper care, they can adapt to both cold and warm climates. A warm, draft-free coop will keep them well during the cold months, while plenty of shade and water help them brace harsh summers. They're generally disease-resistant and able to thrive in most environments. Cleaning the coop regularly and providing them with fresh water will keep pests like mites and lice at bay, as well as decrease respiratory problems.

The New Hampshire Red lay eggs all year round. They produce anywhere between 250 to 300 large brown eggs. While cold weather slows down egg production, it doesn't bring it to a stop. Their calm and friendly disposition makes them easy to handle. Their sociable personalities and non-aggressive nature make them perfect for homesteads, family farms, and beginners. They're active and curious, making them suitable for free-range environments, and surprisingly intelligent.

Cochin

Large and fluffy Cochins are sought after by chicken owners for their friendly nature. The standard-sized Golden Laced Cochin are large, soft, feather-legged chickens highly sought after in the U.S. due to their rare golden laced color. Their dense plumage helps them thrive in cold and moderate climates. A shaded enclosure and access to fresh water can prevent heat stress during the summer months.

Their eye-catching, thick, fluffy feathers require regular grooming to stay in good condition, with their feathered legs and feet requiring special attention. Dirt can cling to their feathers in wet conditions, making regular cleaning and maintenance a must. They're moderate egg layers, producing 180 to 220 eggs a year. The eggs are medium to large-sized and brown in color.

While egg production slows down during winter, lengthening the daylight hours through artificial lighting can make them continue laying consistently. They're gentle and friendly with a calm, docile nature, making them perfect for backyard flocks and for first-time chicken keepers. They interact well with other chickens, existing peacefully in a mixed flock.

Black Australorp

Calm and friendly, these chickens lay a large number of eggs, making them perfect for backyard chicken owners. The sweet-natured hens make excellent pets. They're cold-hardy and great at foraging. The breed was developed in Australia primarily for egg production; however, the birds are prized for their meat as well. They're an excellent choice for the backyard, especially if you want an uninterrupted supply of eggs all through the winter. They're one of the best egg-laying breeds, producing anywhere between 250 to 300 large brown eggs each year. Egg production may slow down during the winter months due to shorter daylight hours, but they continue laying eggs nonetheless.

Although they're low-maintenance, they require shaded areas and a cool living space during the summer months so they don't overheat. They're mostly disease-resistant; however, they can fall victim to common pest infestations like mites and lice. An unclean living space without good ventilation can give rise to respiratory infections. They require a balanced diet, fresh water, and a clean and secure environment to thrive. While their exceptional foraging skills make them well-suited to free-range living, they also do well in confined spaces as long as they're cared for.

Their calm, gentle demeanor makes them one of the most docile chicken breeds. They're an excellent choice for beginners or families with small children. They're friendly, easy to handle, affectionate, and don't make a lot of noise, making them perfect for urban areas.

Brahma

These gentle giants thrive in cold climates. Their stand-out features are their feathered legs and their stunning white plumage laced with black. They lay plenty of large and brown eggs, even during winter, while providing ample meat. Their dense feathers make them well-suited to colder climates. Roosters can reach up to 12 lbs while hens reach up to 10 lbs. They're slow to mature, but their calm and gentle nature makes them a top pick for homesteads and family farms.

The Brahma's bulky bodies and pea-sized combs help them survive cold weather. They're low-maintenance overall but require some attention when it snows or the rain makes the ground muddy due to their feathered feet. They need large nesting boxes and adequate space to move around. They're friendly with other breeds and will make a nice addition to your diverse flock.

French Black Copper Marans

Hailing from the western region of France, the Black Copper Marans lay large, dark chocolate-brown eggs that are the darkest of all the Maran types. The hens thrive in a free-range environment. They mature fast, making them an excellent dual-purpose breed. Roosters can weigh up to 8.5 lbs, while hens weigh 7 lbs. Their superior egg-laying ability makes them the perfect choice for egg enthusiasts. Their gentle nature and dual-purpose capabilities make them ideal for backyard flocks.

While they prefer free-range environments where they can roam and forage, they also do well in large, well-ventilated coops with comfortable rotating space and clean bedding. A balanced diet of high-quality poultry feed, fresh water, vegetables, grains, and mealworms will keep these cluckers happy and healthy.

Black Jersey Giant

Originating from Burlington County, New Jersey, in the 1880s, the Black Jersey Giant can reach an impressive size. A great dual-purpose breed, these birds can weigh up to 14 pounds. They're slow to grow, but eventually catch up with the other breeds, surpassing them in size and egg-laying capacity. Their black feathers develop a distinctive green sheen as they mature.

They can tolerate both hot and cold weather, but like most other chickens, require proper shelter. While their massive size makes them thrive in frigid conditions, they still need a well-ventilated, draft-free coop during plunging temperatures. These robust chickens are mostly disease-resistant but require regular checks for parasites, such as mites and lice. A clean environment and regular health checks are important to prevent health problems these large-sized birds are prone to and ensure their overall well-being.

While being notoriously low-maintenance, these gentle giants require fresh water, a balanced diet, and a clean space with plenty of room to roam around to stay healthy. They're excellent egg layers, consistently laying large brown eggs. Like most hens, their egg laying capacity declines during winter due to short daylight hours. Installing supplemental lighting can help maintain egg production even when the mercury drops.

Now that you've learned about the most common backyard breeds and their traits, the table below offers a simple summary of the key details. Use it as a quick reference to compare egg production, temperament, hardiness, and size so you can easily decide which breeds best fit your needs.

Chicken Breed and Features

Breed	Purpose	Egg color	Eggs laid per year	Temperament	Hardiness	Size
Rhode Island Red	Meat and eggs	Brown	250 to 300	Friendly, docile, good forager	Heat and cold tolerant	Medium and bantam
Plymouth Rock	Meat and eggs	Brown	200	Docile, good forager	Cold-hardy	Medium and bantam
Buff Orpington	Meat and eggs	Brown	110 to 160	Friendly, docile, good forager	Cold-hardy	Large, medium, and bantams
Leghorn	Eggs	White	300	Flighty, good foragers	Heat tolerant	Medium and bantams
Amerecauna	Meat and eggs	Pink, blue, green		Docile, good foragers	Heat and cold tolerant	Medium and bantams
Easter Eggers	Meat and eggs	Pale blue to green		Docile, good foragers	Heat and cold tolerant	Medium and bantams
Wyandotte	Meat and eggs	Brown	200	Docile, good foragers	Cold hardy	Large, medium, and bantams
Silkie	Pets	Small, cream to beige colored	100 to 120	Docile, friendly	Heat tolerant	Bantam
Sussex	Meat and eggs	Brown	280	Docile, good foragers	Cold-hardy	Medium and bantam
New Hampshire Red	Meat and eggs	Brown	200	Mostly docile, good foragers	Heat tolerant	Medium and bantam
Cochin	Pet, meat, and eggs	Brown	150 to 180	Calm, docile, good foragers	Cold-hardy	Large, medium, and bantam
Black Austalop	Meat and eggs	Brown	200 to 250	Mostly docile, good foragers	Cold-hardy	Medium and bantam
Brahma	Meat and eggs	Brown	150 to 200	Docile, friendly	Cold-hardy	Large
French Black Copper Marans	Meat and eggs	Dark chocolate, russet brown	140	Docile, friendly	Cold-hardy	Large and heavy
Black Jersey Giant	Meat and eggs	Large, brown	150 to 200	Docile, friendly	Cold-hardy	Very large

The Takeaway

Choosing the right chicken breed really comes down to your goals and lifestyle. Some breeds, like Leghorns, are egg-laying powerhouses, while others, like Silkies, shine more for their personalities and broodiness than their egg basket. Heavier dual-purpose breeds such as Jersey Giants and Marans can give you both meat and a steady supply of beautiful eggs, while ornamental breeds bring charm and character to your flock. Each breed carries its own unique features. Whether it's temperament, climate tolerance, or the color and number of eggs they provide.

The key is to match the breed to your needs: do you want lots of eggs, friendly pets, striking looks, or hardy birds that can thrive in your climate? By taking the time to consider these traits, you'll end up with a flock that not only meets your expectations but also brings daily joy, whether through a basket of fresh eggs or the simple pleasure of watching happy hens scratch and cluck in your yard.

With your ideal breeds in mind, it's time to think about where they'll live. A sturdy, well-planned coop sets the foundation for a happy, thriving flock.

Chapter 3

A House for Your Hens

The chickens in my backyard do more than just provide my family with fresh eggs every morning. They're a delight to watch and keep me entertained all day with their antics. Just last week, I was talking to a neighbor over the fence and Bella, my New Hampshire Red, kept hopping up on my legs, trying to get my attention. I ignored her at first.

When she wouldn't stop scratching my legs, I leaned down to check what was wrong. The moment I did so, my feisty chicken leapt up, snatched the earphone dangling from my ear and bolted. It seemed that Bella had mistaken my earphones for one giant worm! The other hens saw me running after Bella and decided to join in on the fun. My neighbor laughed her head off as she watched a flock of chickens sprinting after me while I sprinted after Bella.

I can think of numerous such amusing incidents involving my chickens. These friendly, lovable creatures sure know how to make their owners laugh. They're fun, curious, clumsy, and full of surprises. With the little cluckers around, you won't have a single dull moment in your life. So let's look at how to create a home for your feathered friends within your backyard.

House Hunting 101

A clean and spacious coop is the secret to keeping your chickens happy and healthy. I realized the numerous options I had to choose from when I started house hunting for my little cluckers. I had so many questions. Should I build or buy the coop? How many nesting boxes or perches will I need? How much space will it take up? It took me some time to figure out what worked best for my feathered friends. The effort paid off in the end when I had a flock of healthy hens and a basketful of eggs. Choosing the perfect coop requires thoughtful planning and careful consideration. Let's look at some of the factors you should take into account to create the perfect home for your flock.

What Makes the Perfect Coop?

Coops come in various shapes and sizes. The coop you choose depends on numerous factors including whether you plan to keep your chickens in a free-range environment or confined. For instance, if the chickens are allowed to roam around freely all day, then a small pen will suffice, providing them with a warm resting place during the evening. But if you're unable to watch over them or keep them in the open due to predators, then an in-built roaming area within the coop might be a better option.

When it comes to the material of the coop, timber is by far considered the best, but there are numerous other options to choose from. The position of the coop is another important decision you'll have to make. The best option is to choose a place that's partly shaded and partly sunny. If there are no shady areas in your backyard, then you can purchase a removable roof or a coop cover.

While choosing a spot for the coop, take a look around at the surrounding space. If you plan to let your chickens roam around freely, then you want a large-sized space where they'll be safe from predators. It's also best to situate the coop away from water sources to avoid contaminating them with chicken waste. An optimal spot is one that catches the sunlight and allows plenty of space for your little cluckers to wander around.

Another factor to take into consideration is the foundation of your hen house. Large, immovable coops are best placed on a solid flat surface, like a concrete slab. This keeps the flock safe from burrowing predators. Chickens love to roll around in dust or sand. Known as a dust bath, the practice helps them remove parasites and excess oils from their feathers. An area with some sand or loose earth where they can indulge in their beauty routine is essential.

Once you've taken care of the outer details, it's time to turn your attention to the inside of the coop, starting with the bedding. Chickens need a clean, comfortable surface to rest on, and bedding (or litter) is the material that covers the coop floor. It helps absorb excess moisture, keeps odors down, and reduces the risk of disease. Common options include sand, straw, dirt, wood shavings, and hemp.

Buy or Build?

Deciding whether you want to buy or build the coop depends largely on your woodworking skills and your budget. While buying is certainly more convenient, if you're searching for economical options, then it's better to build the coop yourself. Luckily, chickens don't give two hoots about aesthetics. All they need is a safe place where they can rest without being hounded by predators.

So even if you try your hand at building the coop yourself, you don't have to worry about it being perfect. As long as it has a sturdy foundation, stays dry, and provides your hens a safe place to sleep and lay eggs, you're good to go. It's best to aim for a functional setup that you can disassemble when needed.

Cooped Up or Free-Range?

Will you let your flock wander around, pecking the earth in search of delicious worms or keep them confined to the coop? Letting the chickens forage boosts their health and egg production, but there's always the looming threat of predators. You can keep the flock safe by investing in a mobile chicken tractor that offers the birds access to fresh pasture while keeping them safe. Alternatively, you can also confine the hens to a stationary coop with a run.

What About the Manure?

Let's get down to the nitty-gritty. Chicken waste can build up quickly and become a breeding ground for disease if it isn't managed properly. You can handle it in several ways, such as replacing the bedding weekly, using the deep-litter method, or rotating bedding more frequently. The deep-litter method involves letting bedding accumulate over time and adding fresh material on top so the lower layers slowly break down, much like compost, while still keeping the coop clean and warm.

In my experience, the best approach is to use carbon-rich materials for bedding, which can later be transformed into excellent compost. Pine shavings, wood chips, and hemp bedding all provide plenty of carbon. Every few months, I clear out the henhouse and add the used bedding straight to the compost pile.

Don't Forget About the Predators

Your plump, gorgeous hens are bound to draw the attention of the surrounding wildlife. Although the threat of wild predators is significantly low in urban environments, it can't be ruled out entirely. Predator proofing the coop is a must to keep your birds safe and secure. Covering the ground with a bury mesh and hardware cloth and fencing are some ways to keep harmful animals at bay.

How Big Should it Be?

You must provide at least one foot of space per bird to avoid overcrowding your hens. The first coop I bought came with two roosting bars that were 6 feet each. I stuffed it with 14 hens. Within a week the place was overflowing with manure, and I had 14 very angry hens, pecking at each other and making a ruckus. I learned my lesson and built a larger coop that was more to their liking. That helped them calm down and kept the bedding clean and dry for a long time.

The size of your coop depends on the number and breed of chickens in your flock. Small bantams thrive in as little as 2 square feet per bird. Larger hens like the Black Australorp and Brahmas will need a minimum of 10 square feet per chicken. On average, the coop should be big enough to provide three to 5 square feet per standard-sized chicken and contain a nesting box for every three to four hens. If you're still in doubt about the size and have ample space in your backyard, then go for the bigger option. You can also try chicken coop calculators online, like this one: https://www.chickenfans.com/chicken-coop-size-calculator.

Run or No Run?

A run is a fenced-in space attached to the coop for your flock to roam around. You might not need a run if you plan to raise the chickens in a free-range environment. However, if you live in an urban area where you're unable to release your flock outdoors, then adding a run to your coop will allow your chickens to wander under the open sky. Another advantage of having a run attached to the coop is that you can easily lock the hens inside while doing yard work.

Stationary or Mobile?

Stationary coops are permanent structures that you can't move around. You can either purchase them, build them yourself, or convert existing structures like a shed or a barn into one. They tend to be a lot sturdier and provide better protection from fluctuating weather patterns and attack from predators. They're the best choice if you own the property and don't plan to move anytime soon.

A portable or mobile coop, also known as a chicken tractor, allows you to move the flock around. You can easily transfer the coop to another spot, giving your chickens access to a broader area and reducing damage to your yard. These coops are great for relocating chickens to shelter them from bad weather. You can also invest in a semi-mobile coop, which you can convert into a mobile henhouse whenever the need arises.

Types of Chicken Coops

You'll come across various types of coops when you begin searching for the perfect house for your chickens. Let's look at some of these and what they have to offer.

- **Raised Coop:** Like the name suggests, the coop is raised above the ground. They keep the flock safe from the wet, muddy floor, keeping them dry and clean during rain.

- **Chicken Ark:** Also known as an A-frame coop, this is a small setup ideal for bantams and small breeds. It can host two to four hens. It's small and compact so you can easily move it around your yard.

- **Chicken Tractor:** This is a portable coop with wheels that can be moved around.

- **Two Level Coop:** Large-sized coops usually include multiple stories, while providing extra shade under the coop. The elevated design provides extra protection from predators while preventing flooding when it rains.

- **Walk-In Chicken Coop:** These coops are big enough for you to walk inside. They can accommodate 8 to 16 hens.

- **All-in-One Coop:** A single compact unit that includes the coop, run, and nesting boxes. It's convenient for beginners and offers everything your hens need in one structure.

DIY Chicken Coop

Building the coop yourself can be a fun DIY project that cuts the price of your hen house by half. A safe and comfortable home for your chickens is essential for the well-being of your feathered friends. Luckily, you don't have to be a professional woodworker to create a lavish abode for your chickens. Basic DIY skills and a solid chicken coop plan is all you need to build an excellent home for your flock.

Before you start, you must have a clear idea of how much space you have, the size of your flock, and your local climate. The material you choose will also have an immense impact on your chicken coop. Wood remains a popular choice due to its availability and affordability. Once you've answered all the important questions, it's time to gather your supplies and start building. Here's a step-by-step guide to create the perfect chicken coop.

Step 1: Create Your Chicken Coop Plan

The first step is to map out your chicken coop. Decide what kind of structure you want to build, its size and location in your backyard, and the number of nesting boxes required. You need a clear idea of what the coop will look like on the inside and outside. You can find plenty of coop plans online for free as a reference. I found easycoops.com an excellent resource for construction plans for all kinds of coops. Just make sure to keep functionality in mind while selecting a plan. The perfect structure offers easy access inside the coop for cleaning, collecting eggs, and regular maintenance.

Step 2: Gather Supplies

It's time to get your supplies ready. Here's a list of items you'll need

- Building material of your choice (wood/plastic/metal)
- Hardware cloth

- Nails
- Screws
- Drill
- Framing square
- Saw
- Hammer
- Measure tape

If you choose wood as your material of choice, then it should be properly treated to prevent rotting. If you have the budget, then go for higher quality options like redwood, which are resistant to rotting and pest infestation.

Step 3: Finding the Perfect Location

Choose a spot that is safe from predators and allows quick access. If possible, build the coop on high ground to prevent flooding. This will be particularly important if your area regularly receives heavy rains. Keep some distance between the coop and large plants with dense foliage as they make a good hiding place for predators.

Sunlight has a significant impact on egg-laying. Insufficient light can lead to decreased egg production.[8] Position the coop so that it receives maximum southern exposure, ensuring ample sunlight and warmth, while getting some shade throughout the day. This is important because you want to keep your hens cool in hot weather. If there are no shady areas in your backyard, simply covering the run with a tarp will work.

Step 4: Lay the Foundation

Clear any vegetation from your chosen area and level the ground. Use a measuring tape to mark out the coop's footprint. Then dig 8-inch-deep trenches where the walls will sit. Fill the trenches with concrete and allow it to cure before laying bricks on top to create a solid, stable foundation for your coop.

Step 5: Build the Frame

A sturdy, weather-resistant frame will extend the life of your coop. Start by cutting your lumber into 2x4 boards, adjusting the lengths to suit the size of your design. Assemble the base first, then add the vertical posts and the horizontal beams that sit on top. Build the frames for the doors and windows, and finish by installing the supports for the roof.

Step 6: Sheathing the Roof and Walls

Once the frame is complete, it's time to add the sheathing. Plywood is the most common choice because it's sturdy and easy to work with; perfect for beginners. Start by sheathing the walls, then move on to the roof, making sure to cut out the openings for the door and windows as you go.

Step 7: Internal Cladding and Insulation

Select your preferred insulating material: fiberglass, straw bale, cardboard, or wool are all common options. Attach the insulation to the sheathing using staples or glue, then install the internal cladding, which is the layer that covers and protects the insulation on the inside walls. Make sure to trim the cladding for a smooth, clean finish.

Step 8: Build Nesting Boxes

Nesting boxes provide the birds a safe and comfortable place to lay eggs. It should neither be too big nor too small. A standard nesting box measures 16 inch x 16 inch x 16 inch. Make them too big and your chickens may not feel safe and secure. If they're any smaller than the given dimensions, then there may not be enough space for bigger breeds like Orpingtons, Wyandottes, and Jersey Giants. The birds should have easy access to the nesting boxes. The design should make it easier for you to collect the eggs, and there should be plenty of airflow to keep the area fresh and dry.

Nesting box with fake eggs

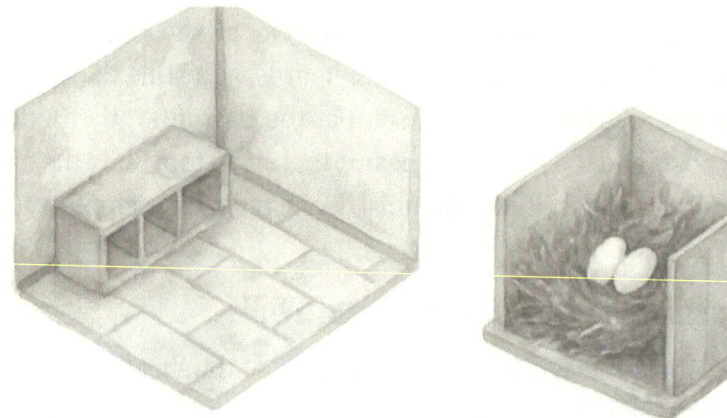

Start by constructing a basic nesting box frame or skeleton as shown in figure 1. Determine the number of boxes you'll need before cutting the wood. You should have at least one box per 3 to 4 hens. Take a flat piece of plywood about 2 inches thick. Measure and cut the side pieces and dividers. You'll need one divider per box plus one extra. For a 16 x 16 x 16 inch box, you'll need 16 inch square pieces.

Next, cut the plywood to create the top, back, and bottom pieces. Multiply the number of boxes you'll need by the width of one box to determine the length of the top and back pieces. The bottom piece should be the same length but at least 10 inches wide to create a small ledge at the front, giving hens a place to step as they enter. Finally, cut a narrow board for the front lip; a short piece that runs across the bottom of each box to keep the eggs and bedding material in place.

Once you have all the pieces ready, it's time to assemble the final structure. Use wood glue to attach the dividers and sides to the frame and secure them with screws or nails. Place the top pieces and fix them in place using 2 inch screws on each side panel and divider. Finish off by attaching the bottom part and the front lip to the bottom front.

Place the finished nesting boxes lower than the roosting bars in the coop but not directly under it. Add soft bedding material like straw, dried grass, and shredded papers to each box. When everything's ready, you can encourage your hens to sit inside the boxes by placing fake eggs or golf balls inside. Hens show broodiness when they settle into a box for long stretches, fluff their feathers, and begin shaping the bedding as though they're preparing to lay an egg.

Step 9: Build Windows and Door

Building the doors and windows involves creating frames, sheathing them with plywood, installing hinges and latches and adding a trim. The design should offer adequate predator protection, withstand strong winds, and heavy downpour while ensuring proper ventilation and easy access.

Build a rectangular frame using 1 x 3 lumber. Join the pieces together with pocket screws and glue. Create a sheathing by attaching plywood to the frame. Place the trim around the edges to prevent drafts and ensure a tight seal. Install the hinges on one side and a handle or latch on the other. Build a square frame for the windows.

Step 10: Install the Roosting Bars/Perch Ladder

Roosting bars or a perch ladder will provide your birds a place to sleep and rest at night. The bars should provide each bird at least 9 to 12 inches of space. They should be placed in a staggered manner at least 12 inches apart, so no chicken sits on top of another and the excrement collects on the ground. Each bar should be 2x2 wide and a minimum of 2 feet above the ground.

The roosting bars should have a smooth finish to prevent injury. They should be made of wood rather than plastic or metal, which can get hot or cold depending on the weather. You can easily construct a perch ladder, which will give your hens a comfortable place to sit. Here are the items you'll need to construct the perfect resting spots for your hens.

- 2x2 pine boards
- Round Dowel Rods
- Fasten brackets
- A box of 3 1/2 inch screws
- A box of 6 inch screws

- 100 grit sandpaper
- Electric drill
- Level
- Ruler
- Miter saw
- Sponge or old rag

Cut the 2x2 boards to make two sides of a ladder. Adjust the length based on how many roosting bars you'll need and the size of your coop. Both sides should be equal in size. Using a pencil and a ruler, mark the places where you'll install the dowel rods. Keep a distance of 12 inches between the rods. The lowest rung should be 2 feet above the ground, and there should be a space of 12 inches between the rod at the very top and the ceiling.

Pre-drill the holes on each board then attach the dowel rods with screws. Use your level and ruler to make sure everything is straight. Once you have the perch ladder ready, use a sandpaper to smooth out the edges and get rid of any rough spots. Lastly, wipe the structure with a damp cloth to remove debris and lean it against the wall inside the coop.

Step 11: Add the Wire Mesh Run Sheathing

Install a secure fence around the run to give your birds a safe place to roam and forage. Use hardware cloth, not chicken wire, which predators can tear through, and attach it firmly to the frame with staples or nails. Make sure the mesh is pulled tight and reaches all the way to the ground. For added protection, bury the bottom edge a few inches deep or bend it outward to create a small predator skirt that helps keep digging animals out.

The Cost of Building A House for Your Hens

Building a cozy abode for your hens costs a fraction of the cost of a ready-made one. The cost of materials ranges from $300 to $1000, depending on the size, design, and materials used. Pre-built coops available on the market range from $1500 to $6000, with the size, design, and number of features driving the cost up. Building the coop yourself gives you more control over the expenditure, allowing you to make adjustments to squeeze the project into your budget. Let's take a closer look at the costs of constructing a house for your hens.

The Frame

The quality and overall cost of the project hinges on the materials you choose. Lumber and plywood are the most popular choices for creating the frame. Let's look at some common options and their costs.

- **Softwood:** It is fairly inexpensive, costing $2 to $3 per square foot.

- **PVC:** Long-lasting, durable, it's used mostly as trim. The cost ranges between $5 to $6 per square foot.

- **Pressure-treated Lumber:** A board foot of pressure treated lumber costs $7 to $10. Its rot and pest-resistant properties make it a popular choice for chicken coops.

- **Redwood:** Rot and pest resistant, it is a relatively expensive option, costing $10 to $12 per board foot.

- **Plywood:** This versatile wood type can be used to construct the walls, roofing, or flooring. It will cost you $10 to $25 per sheet.

- **Corrugated Metal:** Best for walls and the roof, it will cost around $15 to $20 per sheet.

- **Plastic Sheet:** Easy-to-use and malleable, it is a somewhat expensive option, with a single sheet costing around $25 to $30.

Flooring

Great flooring helps deter predators, makes cleaning easier, and provides insulation. Here are some commonly used flooring materials along with their typical price ranges.

- **Vinyl:** Cheap, easy-to-clean and install, this flooring material works well with plywood underneath. It costs $2 to $8 per square foot. The only downside is that it is not as durable as the other options and prone to becoming damaged by moisture and pests.

- **Wooden Decking**: A durable and easy-to-install option that will set you back a $1.40 per foot.

- **Plywood:** Available at $10 to $25 per sheet, this is a popular choice for flooring. However, it's not predator safe and tends to retain odors. An exterior marine grade varnish on both sides can make it more weather-resistant.

- **Wire:** Wire mesh flooring is an excellent way to deter pests. Costing around $40 per roll, it's fairly easy to install but requires a layer of sand, sawdust, or straw on top to make it easier for chicken feet.

- **Rubber Mat:** Comfortable to walk on and excellent at warding off predators, rubber mats will set you back about $45 to $80.

- **Concrete:** Long-lasting, effective at deterring pests, and easy-to-clean, a concrete floor will cost you around $75 to $125 per cubic yard. You'll need a minimum of two coats of clear masonry sealer to prevent odors, stains, and prevent cracking.

A sturdy, well-built coop protects your feathered friends from fluctuating weather and harmful predators. It creates a safe and healthy space for your chickens to lay eggs and rest. The material you use has a direct impact on the quality and cost of a DIY chicken coop as does its design. So let's look at different coop types and how much they'd cost you to construct.

- **Chicken Ark or A-Frame:** It is the most inexpensive and simple design to build that will cost you around $300 to $500.

- **Chicken Tractor:** They're relatively inexpensive and fairly easy to build. The total cost of building a chicken tractor hovers around $500 to $700.

- **Walk-In Chicken Coop:** Constructing one from scratch will set you back a hefty $1000 to $1500 or more depending on its size. You can cut costs by repurposing an old storage shed or other outbuildings.

- **All-In-One Chicken Coop:** You'll have to stretch your budget to up to $2000 to include this coop design. Depending on the size, your budget can reach up to $4000.

The best thing about building a chicken coop yourself is it gives you the ability to control your budget. You can make adjustments to your design or the materials used to squeeze the overall cost within your budget. As a beginner, it's best to start small with no more than 3 or 4 chickens and a coop big enough to house them.

The Takeaway

A well-constructed and comfortable hen house is essential to keep your flock secure and healthy. It's important to familiarize yourself with different coop features before you go out to purchase a pre-made structure or build one yourself. On average, it costs around $300 to $1000 to build a nice little home for your hens. The more features you add, the more the cost goes up. It's best to set a budget and choose materials that fall in your price range.

Once the coop is in place, your focus shifts to caring for the space and the flock that calls it home, keeping everything clean, safe, and well maintained so your hens stay happy and productive.

Chapter 4

Coop Care: Chicken and Egg

Eggs, eggs, and more eggs! With a flock of five hens and a rooster, I end up with way more eggs than I need. Each morning, I put some of the eggs into a basket, hook it onto my arm, and go around the neighborhood, distributing my nature's bounty. My nextdoor neighbor, Judy, loves the daily delivery of fresh eggs. She insists she won't ever buy from the store. Each morning, I find her sitting on her porch with a basket of rhubarb, turnips, and radishes from her vegetable patch along with a few kitchen scraps for my feathered friends, which she'd exchange for the eggs.

When I decided to get chickens, I didn't intend to sell eggs to my neighbors. A few months later, with a grown up flock of five buckeye hens, I had more eggs than I knew what to do with! I decided to distribute them all over the neighborhood. A week later, my neighbors started dropping by my house asking for more eggs. When I told them I didn't have any extra eggs, they insisted on paying me for them! Judy said she couldn't possibly go back to buying eggs from the grocers after tasting the ones laid by my homegrown chickens.

Meeting Nutritional Needs

When it comes to raising happy, healthy chickens, nothing matters more than what goes into their little beaks. Just like us, chickens thrive when they get the right balance of nutrients, enough protein for building strong feathers and muscles, vitamins and minerals for good health, and of course, clean water to wash it all down. While chickens might look like they'll eat just about anything (and honestly, they'll try!), not everything is good for them. That's why understanding their nutritional needs is the foundation of good flock care.

Think of chicken feed like a puzzle: every piece: protein, grains, greens, grit, vitamins, calcium, and water, fits together to keep your birds thriving. Miss a piece, and problems start to pop up: poor egg production, weak shells, slow growth, or health issues. But once you learn the basics, feeding chickens becomes second nature. Let's walk through the essentials, from what goes in commercial feed to the extras you can (and shouldn't) give your flock.

What Does a Chicken Need?

In the wild, chickens are natural foragers. Their ancestors scratched around in forests, eating bugs, seeds, small plants, and the occasional lizard or frog. While today's backyard hens don't need to fend for themselves quite as much, their bodies are built with the same needs in mind. Here's what makes up a balanced diet for your flock:

- **Protein:** This is the building block for feathers, muscles, and eggs. Young chicks especially need high levels of protein (around 18–20%)[9] to grow strong, while laying hens need enough protein to keep those eggs coming regularly.

- **Grains:** Chickens get their energy mostly from grains like corn, wheat, and barley. Think of grains as their "carbs" or the fuel that keeps them scratching and clucking all day.

- **Greens:** Leafy plants and weeds add variety, fiber, and micronutrients to their diet. If you've ever tossed a handful of kale or spinach into the run, you know how fast chickens will pounce on it.

- **Calcium:** Essential for egg-laying hens. Without it, shells turn thin and fragile, and hens can become weak. Oyster shell is the most common supplement, but crushed eggshells also work.[9]

- **Vitamins:** Vitamin A supports growth and immunity, while Vitamin D helps hens absorb calcium for strong shells.

- **Insoluble Grit:** Chickens don't have teeth, so grit (tiny stones) is nature's way of helping them grind up food in their gizzard.

- **Water:** The most overlooked nutrient! A chicken's body is mostly water, and without enough, they won't eat, lay eggs, or stay healthy.

If you provide these basics in the right amounts, your chickens will reward you with good health, happy clucks, and plenty of eggs.

Complete Feed vs. Scratch and Grain Feeding

One of the biggest questions new chicken owners ask is: Do I really need to buy complete feed, or can I just give my chickens grains like corn and wheat? The short answer: yes, you do need complete feed.

Complete feed is specially formulated to give chickens all the nutrients they need in one package. It's like their daily multivitamin, balanced to the last detail. Feeding only grains, on the other hand, is like giving kids candy for every meal. Sure, they'll be full but they'll be missing crucial vitamins, minerals, and protein.

Scratch grains, a mix of cracked corn and other grains, are more like a treat. Chickens love scratching around for them, and they're great for entertainment, but they should only make up about 10% of the diet. Think of scratch as dessert, not dinner.

Feeding Chickens by Age and Type

Not all chickens have the same dietary needs. Just like toddlers, teens, and adults need different foods, so do chicks, pullets, and layers. Here's the breakdown:

- **Chicks (0–6 weeks):** High-protein chick starter (18–20%), often medicated to prevent coccidiosis (a common disease in young birds).[10]
- **Pullets (6–20 weeks):** Grower feed with slightly less protein (14–16%), no added calcium yet. They're not laying, so too much calcium at this stage can actually harm them.
- **Laying Hens (20+ weeks):** Layer feed with about 16% protein and added calcium. This is the sweet spot for keeping eggs coming with strong shells.
- **Brooders and Roosters:** A balanced feed without excess calcium, since roosters don't need strong eggshells. Too much calcium can damage their kidneys.
- **Meat Birds:** These chickens grow quickly and need a high-protein (20–22%) broiler feed to reach market weight in a matter of weeks.
- **Dual-Purpose Breeds:** Heavier egg layers like Rhode Island Reds and Orpingtons do well on regular layer feed, but they may need extra protein during molting.

A good trick is to always check the feed bag label. It will tell you if the feed is meant for chicks, growers, or layers.

Calcium and Grit: The Unsung Heroes

If you've ever cracked open a thin, soft-shelled egg, you've seen what happens when hens don't get enough calcium. Laying hens should always have free-choice access to calcium, like crushed oyster shells in a separate dish. That way, they can take it when they need it.

Grit, on the other hand, is essential for digestion[11]. If your chickens free-range, they'll often find their own. But if they're confined or on softer soil, you'll want to provide grit to help them break down feed properly. Without it, they may suffer from impacted crops or poor digestion.

The Best Form: Mash, Pellets, or Crumbles?

Chicken feed comes in a few different forms, each with its own pros and cons. Mash is a fine, powdery version that's inexpensive but can be messy since chickens often scatter and waste it. Pellets, on the other hand, are compact little nuggets that reduce waste, are easy to handle, and guarantee that every bite is balanced with the right nutrients. Crumbles fall somewhere in between. They're simply broken-down pellets, which makes them especially good for younger birds that may struggle with larger pieces. For most backyard keepers, pellets or crumbles end up being the preferred choice because they keep things tidy and help minimize waste.

Feeding Behavior and Pecking Order

Chickens are flock animals, and feeding time is often when their personalities really show. The dominant hens, the "top of the pecking order" eat first, while timid ones may hang back. To make sure everyone gets enough, provide multiple feeders and spread them out. A general rule is one feeder per 5–6 birds.

Supplements, Treats, and Foraging

Here's the fun part: what extras can you give your chickens? While their main diet should always be balanced feed, a few extras are fine and even beneficial.

Safe treats: Fruits (like apples, berries, melons), veggies (carrots, cucumbers, leafy greens), and cooked grains.

Natural supplements: Herbs like oregano (natural antibiotic)[12], garlic (immune support)[13], and apple cider vinegar (a digestive tonic)[14] are popular among chicken keepers.

Probiotics: Helpful for gut health, especially after antibiotic treatments.

But be careful. Some foods are toxic to chickens, including chocolate, avocado, raw beans, onions, and anything moldy. Always double-check before tossing scraps their way. As for foraging, free-range chickens will happily gobble bugs, weeds, and grass. While this adds variety and nutrition, it's not enough on its own. Free-range birds still need a base of complete feed.

Feeding Tips and Tricks

A few practical things make a big difference. Store feed in sealed containers to keep out moisture, rodents, and pests. Don't overfill feeders. Chickens waste less if feed is topped up daily. Always provide clean, fresh water. Chickens will stop eating if they don't have enough water. Monitor your flock. If they're leaving certain feed behind or picking out favorites, consider switching feed type or form.

A balanced diet is the secret to healthy, productive chickens. While they love scraps, treats, and foraging, their main diet should always be a complete feed appropriate to their age and purpose. Add in calcium, grit, water, and the occasional safe treat, and you'll have a thriving flock.

Maintaining Proper Hydration

If feed is the "fuel" for your chickens, then water is the oil that keeps their engines running smoothly. You can have the best feed in the world, but if your chickens don't have enough water, they simply won't thrive. In fact, water is so important that a hen will stop laying eggs if she doesn't drink enough, even if she's getting all the nutrients she needs from her feed. It's easy to underestimate water needs because chickens seem so small, but you'll be surprised by just how much they can drink in a day.

Water is the quiet hero of a chicken's health, playing a role in almost every function of their body. Without it, digestion slows down because feed can't be softened and broken apart properly. Chickens also rely on water to regulate their temperature, since they don't sweat and instead cool themselves through panting and drinking.

For laying hens, water is even more critical, after all, an egg is about 75% water[15], so dehydration is one of the first reasons a hen might stop laying. On top of that, nutrients from protein, vitamins, and minerals can't be absorbed efficiently without water as the carrier. In short, when water intake drops, the effects ripple through every part of a chicken's health, from growth to egg production.

On average, a full-grown laying hen will drink about half a liter of water per day under normal conditions[16]. That's roughly two cups of water per bird. But this isn't a fixed number and varies depending on the weather, diet, and whether they're laying. Hot weather can easily double or even triple water consumption. If you've ever peeked at your waterer on a hot summer afternoon, you'll notice it drops fast!

A good rule of thumb: always provide more water than you think they'll need. Chickens aren't shy about drinking frequently, and keeping their supply topped up ensures they never run dry.

Food Quality Control: Maintaining Food and Water Hygiene for Healthy Chickens

Even with high-quality ingredients, poor handling and storage allows bacteria, molds, dust, soil, and heavy metals to infiltrate the feed, affecting its nutritional value. Contamination diminishes vitamins like A, D_3, E, K, and B_1,[17] undermining flock performance.

To prevent this, it's important to store the feed in cool, dry, pest-proof containers. Sealed bins keep out moisture and rodents. You can also decrease chances of contamination by following rigorous hygiene protocols while handling or preparing the feed.

Disease-causing microorganisms include *Clostridium perfringens*, which causes necrotic enteritis, a serious intestinal infection that leads to tissue damage and reduced nutrient absorption; *Salmonella,* a risk to birds and humans alike; *Escherichia coli,* especially dangerous to young chicks; and molds, which reduce feed intake and may produce mycotoxins or respiratory irritants[18]. Using feed and water-safe additives, such as a commercially available hygiene liquid, can also inhibit bacterial and mold growth.

Water contamination is a particular cause of concern as it spreads rapidly, especially during warm weather. Standing water exposed to the environment can quickly become a breeding ground for E. coli, Salmonella, algae, and other pathogens. Even well-designed drinkers must be cleaned frequently and periodically dismantled for deep cleaning.

Rotate and discard any feed giving off bad odors, showing clumps, or signs of mold. Clean feeders between batches, and flush and refill drinkers daily. Combine these habits with a suitable hygiene additive and you'll drastically lower contamination risk. The payoff: healthier birds, fewer losses, more productivity, and peace of mind that your flock is drinking and eating safely.

Eggs in Your Basket

There's nothing quite like the excitement of discovering that first warm egg nestled in the nesting box. It marks the moment your patience, care, and good feed finally pay off. Most pullets begin laying around six months of age, though timing can vary widely. Some early bloomers may surprise you with their first eggs at just 16–18 weeks, while others take their time, beginning closer to 28–32 weeks. On average, 20–22 weeks is when most backyard keepers see the first egg appear.[19]

Breed plays a big role, too. Chickens bred for high egg production, such as Leghorns, Golden Comets, Rhode Island Reds, and Australorps, usually mature faster and start laying sooner. Heavier or more ornamental breeds like Orpingtons, Wyandottes, and Barred Rocks tend to take a little longer. Easter Eggers, known for their beautiful blue and green eggs, are often the last to join the laying club[19].

If your chicks mature in autumn or winter, shorter daylight hours might delay egg-laying until spring. It's best to let nature take its course; hens need rest during the colder months, and their bodies know when it's time to pause and when to begin again.

Five Signs Your Hens Are About to Lay Eggs

After months of feeding, watching, and waiting, your young hens are nearly ready to reward you with their first eggs! Here are five telltale signs that "egg day" is just around the corner.

1. Bright Red Combs and Wattles

As a pullet matures, her comb and wattles, once pale and small, begin to swell and turn a deeper red. This color change signals that her hormones are shifting and her body is preparing to lay. If this happens very early (before 8 weeks), it might mean you have a young rooster instead!

2. Exploring the Nesting Boxes

A curious hen will start spending more time in or around the nesting boxes. She may even sit inside as if testing them out. To encourage her to lay in the right spot, place fake eggs or even golf balls in the boxes. Hens prefer laying where other "eggs" already are.

3. Becoming More Vocal

You might hear more chatter as hens approach laying age. Some even perform an egg song: a mix of squawks and clucks before and after laying. It's their way of announcing that something exciting is happening!

4. Bigger Appetite

Producing eggs requires energy and nutrients. As hens prepare to lay, they'll eat more. Around 18 weeks, start transitioning them to a layer feed rich in calcium for strong shells. You can also enrich the feed with crushed egg shells to increase their calcium intake.

5. The Submissive Squat

The clearest sign that your hen is ready to start laying eggs is that she crouches low with wings slightly out when you come near. This is known as the "submissive squat." The behavior shows that she's ready to mate and will most likely start laying eggs within a few days.

The Secret Life of Farm-Fresh Eggs

Eggs straight from the coop are a different kind of magic. Not only do they taste better and often display beautiful natural colors, but they also arrive untouched by industrial processing. That means they need a bit more care in storage and cleaning than supermarket eggs.

To Refrigerate or Not?

You can safely keep unwashed eggs at room temperature for a couple weeks without worry. Eggs are covered with a protective coating called bloom. It fills the tiny, microscopic pores on the shells, keeping out bacteria and other harmful substances. Washing removes this natural barrier, making it more susceptible to bacterial contamination and making refrigeration essential. Once an egg has been chilled, it should stay chilled. Leaving it out at room temperature for too long can cause it to spoil.

It's best to avoid washing eggs until you actually need to use them. If an egg is obviously dirty, use warm water and give it a mild rinse. Use a soft cloth or sponge to remove dirt or droppings and air-dry before storing. Eggs stored in the fridge last far longer. Unwashed, "bloom-on" eggs stay good for two to three months; washed eggs lose moisture more quickly and should be stored in covered containers to protect them from odors and bacterial contact[20].

Test Their Freshness

The float test is a classic trick to check the freshness of eggs. Fill a bowl with cool water and gently slide in the egg. Fresh eggs will sink flat to the bottom. As they age, they'll begin to tilt or even float. The more they float, the more air has seeped in over time, indicating their age.

EGG FLOAT FRESHNESS TEST

VERY FRESH, AND GOOD TO EAT   **STILL FRESH, USE IT SOON**   **TOO OLD, DISCARD THESE**

Keeping the Coop Clean

Much of preserving egg quality happens before you even bring an egg inside. Keep nesting boxes tidy, change soiled bedding promptly, and discourage hens from sleeping or perching inside nests. A clean, dry nest reduces dirt and contamination. Place dummy eggs in the boxes to teach hens where to lay (but don't leave too many as it can trigger broodiness).

When Chickens Turn Snackers (And How to Stop It)

Imagine heading out to the coop one morning, basket in hand, expecting a fresh batch of eggs only to find cracked shells. Unfortunately, egg eating happens, and once a hen figures out how delicious her own eggs are, it becomes a habit that's hard to break. But don't despair! With a few smart tips and tricks, you can keep your flock from turning into egg-eaters.

Why It Happens

Egg eating isn't a chicken's default behavior. Instead, it's usually triggered by something external. If your hens are craving extra protein or calcium, they may resort to eating eggs to fill those needs. Bored chickens, especially during winter or when stuck indoors, sometimes peck their own eggs out of curiosity or frustration. Once one hen cracks open an egg, others may follow her lead, thinking it's a snack that's fair game.

Preventing the Problem Before It Starts

Never let eggs pile up in the nesting boxes. Collect them once or twice daily so there's less temptation for your hens to break one open. Make sure the nest boxes are roomy, cozy, and cushioned (think two inches of soft bedding). If you have too few boxes, hens will overcrowd them, increasing the risk of crushing eggs.

Ensure each hen has proper nutrition, too. A well-balanced layer feed with enough protein and calcium is essential. Offering free-choice oyster shells or crushed, dried eggshells helps satisfy any extra calcium cravings. Just make sure they're crushed extremely well, so the hens don't confuse them with their own eggs. Treats like black soldier fly grubs can also boost protein.

Breaking the Bad Habit

Remove all raw-egg treats so hens don't associate egg with snack time. Introduce dummy eggs (or even golf balls) in the nest so when a curious hen pecks, nothing cracks open. In extreme cases, you can use sloping nesting boxes in which the egg rolls out of reach of the chickens the instant they lay it. Keep the flock busy and content. Grazing boxes, fresh litter, and plenty of space do wonders to reduce boredom and with it, bad behavior. A blend of good diet, regular egg collection, comfy nests, and creative deterrents can help you nip egg eating in the bud, saving both your eggs and your sanity.

The Joy of Happy Hens: Fun Enrichment Ideas

Your hens aren't just egg-laying machines, they're curious, social, and surprisingly intelligent creatures. When they have things to peck, scratch, and explore, they're much happier, healthier, and far less likely to develop bad habits like feather pecking or egg eating. Enrichment keeps your flock's minds active and their bodies moving, turning the coop into a lively, stimulating space instead of a dull pen. Here are some simple yet entertaining ideas to keep your hens engaged:

- **Hang snacks:** Suspend cabbages, lettuce heads, or corn cobs so the hens can peck and chase them. It mimics their natural foraging behavior.

- **Offer dust baths:** Provide a shallow box filled with dry sand, soil, and a bit of wood ash. Dust bathing helps chickens stay clean and free from parasites.

- **Add toys and mirrors:** Chicken swings, perches, and small mirrors give your birds something to interact with and explore during the day.

- **Encourage foraging:** Toss leaves, hay, or a handful of mixed grains across the run to keep them busy scratching and searching.

- **Rotate activities:** Switch out toys or change the location of treats every few days to keep things interesting.

Egg Abnormalities Guide: Weird Eggs 101

The first time I found a wrinkled, misshapen egg in my nesting box, I panicked. Was my hen sick? Dying? Had I done something terribly wrong? I frantically Googled "weird chicken egg" and spent the next hour spiraling through forum posts about diseases and deficiencies.

Turns out, that weird egg was just... a weird egg. It happens. Chickens aren't perfect little egg factories, and occasionally their bodies produce something unexpected. Most egg abnormalities are harmless, temporary, and downright fascinating once you understand what causes them.

Let's decode the most common strange eggs you'll encounter and what (if anything) you should do about them.

1. Soft-Shelled or Shell-Less Eggs

What It Looks Like:

An egg with no hard shell, just a thin, rubbery membrane holding the contents together. Sometimes the membrane is intact; other times it breaks, leaving a squishy mess in the nest.

What Causes It:

- **Calcium deficiency:** The most common cause. Hens need lots of calcium to produce hard shells.
- **Vitamin D deficiency:** Without enough Vitamin D, hens can't absorb calcium properly.
- **Stress or illness:** Disease, heat stress, or sudden fright can disrupt shell formation.
- **Immature reproductive system:** Young pullets just starting to lay often produce shell-less eggs as their bodies adjust.
- **Overly rapid egg production:** Sometimes a hen's body doesn't give the shell gland enough time to do its job.

What to Do:

- Offer free-choice **oyster shell or crushed eggshells** so hens can self-regulate calcium intake.
- Ensure your flock gets **adequate sunlight or Vitamin D** supplements (especially in winter).
- Switch to a high-quality layer feed with **proper calcium** content (3.5-4%).
- If it continues beyond a week or two, check for illness or reproductive issues.

Is It Safe to Eat?

Yes, as long as the membrane is intact and the egg is fresh. Just handle it carefully!

2. Double Yolk Eggs

What It Looks Like:

Crack it open, and surprise, two yolks instead of one! These eggs are often larger than normal.

What Causes It:

A hen releases two yolks at the same time, and both get encased in a single shell. This is most common in:

- Young pullets whose reproductive systems are still regulating
- High-production breeds like Leghorns or Rhode Island Reds
- Hens in peak laying condition

What to Do:

Nothing! Double yolkers are completely normal and usually phase out as the hen matures.

Is It Safe to Eat?

Absolutely. In fact, many people consider them a treat!

Fun Fact: Triple and even quadruple yolk eggs exist, though they're extremely rare.

3. Blood Spots or Meat Spots

What It Looks Like:

A small red or brown spot on the yolk or in the egg white.

What Causes It:

- **Blood spots:** Tiny blood vessels in the hen's reproductive tract rupture during ovulation, releasing a drop of blood into the egg.
- **Meat spots:** Small pieces of tissue from the oviduct slough off and end up in the egg.

What to Do:

Nothing. Blood and meat spots are harmless and occur naturally. They're more visible in fresh eggs because the whites are clearer.

Is It Safe to Eat?

Yes! You can remove the spot with a spoon if it bothers you, or just eat it; it's perfectly safe.

Cultural Note: In some countries, blood spots are considered signs of a fertilized egg, but this isn't true. Blood spots happen in both fertile and infertile eggs.

4. Wrinkled or Bumpy Shells

What It Looks Like:

Instead of a smooth surface, the shell has ridges, bumps, wrinkles, or a rough texture.

What Causes It:

- **Stress:** Sudden disturbances (predators, loud noises, rough handling) during shell formation.
- **Illness:** Respiratory infections like infectious bronchitis can damage the shell gland.
- **Calcium imbalance:** Too much or too little calcium affects shell quality.
- **Overcrowding:** Stressed hens in cramped conditions produce irregular eggs.
- **Age:** Older hens sometimes lay wrinkly eggs as their reproductive systems slow down.

What to Do:

- Reduce stressors in the coop.
- Check for respiratory illness (coughing, sneezing, nasal discharge).
- Ensure proper calcium and Vitamin D intake.
- Improve coop conditions (space, ventilation, cleanliness).

Is It Safe to Eat?

Yes, though wrinkled eggs may have thinner shells and shorter shelf lives.

5. Tiny "Fairy Eggs" or "Fart Eggs"

What It Looks Like:

A miniature egg, sometimes the size of a marble, often with no yolk inside.

What Causes It:

These occur when a hen's reproductive system misfires, releasing a bit of tissue or debris instead of a yolk. The oviduct treats it like an egg and wraps it in albumen and a shell. Common in:

- Young pullets just starting to lay
- Older hens nearing the end of their laying years
- Hens experiencing hormonal disruptions

What to Do:

Nothing. Fairy eggs are quirky but harmless. Enjoy the novelty!

Is It Safe to Eat?

If there's anything inside, yes, but most fairy eggs contain only egg white or nothing at all.

6. Lash Eggs (Salpingitis)

What It Looks Like:

A rubbery, misshapen mass that looks more like a chunk of tissue than an egg. It may contain pus, blood, or foul-smelling material.

What Causes It:

Lash eggs are a serious sign of **salpingitis** (infection and inflammation of the oviduct). This condition is caused by bacterial infections, often *E. coli* or *Salmonella*.

What to Do:

This requires immediate veterinary attention. Salpingitis can be fatal if untreated. Symptoms include:

- Lethargy
- Loss of appetite
- Swollen abdomen
- Difficulty walking (penguin stance)
- Discharge from the vent

Treatment typically involves antibiotics and supportive care.

Is It Safe to Eat?

No. Discard lash eggs and any eggs laid by an infected hen until she's fully recovered and cleared by a vet.

7. Thin or Translucent Shells

What It Looks Like:

A shell so thin you can almost see through it, or one that cracks easily under gentle pressure.

What Causes It:

- Calcium or Vitamin D deficiency
- Heat stress (reduces calcium absorption)
- Illness affecting the shell gland
- Old age (hens lay progressively thinner shells over time)

What to Do:

Boost calcium intake, ensure adequate Vitamin D, and check for illness. If thin shells persist, consult a vet.

Is It Safe to Eat?

Yes, but handle carefully, they break easily!

8. Flat-Sided or Torpedo-Shaped Eggs

What It Looks Like:

An egg with one flat side, or an elongated, torpedo-like shape instead of the usual oval.

What Causes It:

- Pressure in the oviduct from another egg forming too quickly
- Immature reproductive system in young layers
- Oviduct abnormalities or damage

What to Do:

Usually nothing, these eggs are harmless oddities. If they occur frequently, check for illness or injury.

Is It Safe to Eat?

Yes!

9. Speckled or Spotted Shells

What It Looks Like:

Dark speckles or spots on the shell, often brown or reddish.

What Causes It:

Excess calcium deposits or pigment irregularities during shell formation. Sometimes caused by:

- Stress or disturbances during laying
- Dietary changes
- Normal variation (some hens just lay speckled eggs!)

What to Do:

Nothing, speckled eggs are beautiful and perfectly normal.

Is It Safe to Eat?

Absolutely!

10. Eggs with Ridges or "Checks"

What It Looks Like:

A thin, raised line or ridge circling the egg, or small cracks that have been repaired with extra shell material (called "body checks").

What Causes It:

The egg gets slightly damaged in the shell gland (from stress, injury, or pressure), and the hen's body repairs it by adding an extra layer of calcium.

What to Do:

Nothing, though frequent body checks may indicate stress or illness.

Is It Safe to Eat?

Yes, but these eggs may spoil faster due to compromised shell integrity.

When to Worry

Most egg abnormalities are temporary and harmless. However, contact a vet if you notice:

- Persistent soft shells (more than a week despite calcium supplementation)
- Lash eggs or foul-smelling discharge
- Sudden drop in egg production combined with abnormal eggs
- Lethargy, loss of appetite, or swollen abdomen in the hen
- Blood in the nest or on eggs (sign of prolapse or internal injury)

Quick Reference: Egg Abnormality Chart

Abnormality	Cause	Action Needed?
Soft/shell-less	Calcium or Vitamin D deficiency	Offer oyster shell; check diet
Double yolk	Normal in young/high-production hens	None
Blood/meat spots	Ruptured blood vessel; tissue sloughing	None (safe to eat)
Wrinkled shell	Stress, illness, calcium imbalance	Reduce stress; check for illness
Fairy egg	Hormonal misfire (common in young/old hens)	None
Lash egg	Salpingitis (oviduct infection)	VET IMMEDIATELY
Thin shell	Calcium/Vitamin D deficiency; age	Supplement calcium; check health
Flat-sided / torpedo	Oviduct pressure; immature system	None (usually temporary)
Speckled shell	Pigment/calcium irregularity	None
Ridge/body check	Minor damage repaired by hen	None

Weird eggs are part of the backyard chicken experience. Most are harmless quirks; your hen's way of keeping life interesting. Keep calcium levels up, minimize stress, and monitor your flock's health, and you'll navigate egg oddities like a pro.

And hey, when you crack open that double-yolker or spot a fairy egg in the nest, take a moment to appreciate the weirdness. It's all part of the egg-venture!

Seasonal Chicken Care

From sweltering summers to frosty winters, your flock's needs change with the changing weather. Small changes can go a long way to keep your girls happy, healthy, and laying no matter what nature throws their way.

Summer Care

Chickens don't sweat, so keeping them cool during the summer comes down to smart coop management. A cool shaded area can prevent heat stress during the harsh summer months. Ventilation in the coop is another must to keep the temperature bearable. You can also use shallow pans of water so they can cool their feet, and always make sure their drinking water is fresh, clean, and cool.

Summer heat can take a toll on your flock's appetite and egg production, so a little extra attention can make a big difference. Offer hydrating treats like chilled watermelon, cucumber slices, or frozen berries. These not only cool your hens down but also keep them entertained. Feeding during the cooler parts of the day, such as early morning or evening, helps them eat comfortably without overheating. If possible, sprinkle a light mist of water around the run (but not directly on the birds) to bring down the temperature and create a more comfortable environment.

It's also wise to keep an eye out for signs of heat stress like panting, holding wings away from the body, and lethargy are all red flags. A simple way to help is by creating cross-breezes in the coop using safe fans or vents positioned away from direct drafts. Provide plenty of shade by hanging cloth tarps or planting fast-growing shrubs nearby. Remember, prevention is always easier than recovery. Once a bird becomes overheated, it takes time and care to bring her temperature down. With consistent attention and a few clever adjustments, your flock can stay happy, healthy, and productive even through the hottest summer days.

Winter Care

When the days grow short and icy winds roll in, you can keep the coop nice and warm with a few smart adjustments. Winterizing your coop includes installing wind blocks, proper ventilation, and snug insulation. Good ventilation is just as important in winter as in summer. It prevents moisture buildup, which causes frostbite on combs and wattles (not the cold itself!). Tarps, plastic sheeting, or well-placed walls can block drafts without trapping damp air.

Use deep, insulating bedding like hemp or straw to keep the floor warm and cozy. For waterers, invest in a heater or change water frequently to keep it from freezing. Unless you live somewhere truly frigid, skip the heat lamps as they're a fire hazard and can make your chickens less resilient to temperature swings.

Your hens will lay fewer eggs in winter, but don't panic as it's their natural rest period. If you'd like to gently boost production, try supplemental lighting that mimics daylight hours (but never 24/7 light, even chickens need beauty sleep).

Lastly, don't forget nutrition. Offer scratch grains in the evening to help them generate warmth overnight, and toss in the occasional treat like cabbage heads or winter greens to keep them strong and healthy.

Whether your flock is sunbathing in July or fluffing up in January, the key to seasonal chicken care is observation and balance. Watch their behavior, listen to their clucks, and make small adjustments before problems arise. A shaded area here, a windbreak there, makes a significant difference in keeping your flock safe and sound through the shifting seasons. Caring for chickens is really about partnership: they give you eggs, company, and endless amusement, and you give them comfort through every season.

Winter Egg Production

Winter Laying: Managing Expectations and Maximizing Production

It was mid-January, and I found myself standing in front of my nesting boxes, staring at a grand total of two eggs. Two. From a flock of eight hens. I'd been spoiled by summer's overflowing baskets, and now I felt personally betrayed by my feathered ladies who were contentedly fluffed up on their roosts, seemingly retired from their egg-laying careers.

That's when I learned the truth: chickens aren't egg machines, and winter changes everything.

Why Egg Production Drops in Winter

Chickens are incredibly sensitive to daylight. Their bodies use light as a signal to regulate hormones, particularly those controlling egg production. During spring and summer, long days (14-16 hours of light) keep their reproductive systems in high gear. But as fall arrives and days shorten, that signal weakens. By winter, with only 8-10 hours of natural daylight, many hens slow down or stop laying entirely.

This isn't laziness, it's biology. In the wild, chickens wouldn't waste precious energy producing eggs during a season when food is scarce and chicks would struggle to survive. Instead, they conserve energy to stay warm and healthy.

What to Expect in Winter

Every flock and every hen is different, but here's a general idea of what winter brings:

- **Reduced Frequency:** Hens that laid 5-6 eggs per week in summer may drop to 2-3 per week in winter, or stop entirely.
- **Smaller Eggs:** Winter eggs are often slightly smaller as hens prioritize survival over production.
- **Irregular Timing:** Eggs may come at odd hours instead of the predictable morning routine.
- **Breed Matters:** Cold-hardy, production-focused breeds like Rhode Island Reds, Australorps, and Barred Rocks tend to keep laying (albeit less) through winter. Lighter, flightier breeds like Leghorns and ornamental breeds like Silkies often stop completely.
- **Age Matters:** Young hens in their first year may continue laying longer into winter than older hens who've been through multiple seasons.

To Light or Not to Light? The Great Debate

Many backyard chicken keepers face this question: should I add supplemental lighting to maintain egg production through winter?

Here's what supplemental lighting does: by extending "daylight" hours artificially (using a timer and coop light), you can trick your hens' bodies into thinking it's still spring or summer, which keeps them laying. Sounds great, right?

The Pros:

- Consistent egg production year-round
- No awkward winter "egg drought"

- Useful if you sell eggs or rely on them for income

The Cons:

- **Shortened lifespan:** Hens are born with a finite number of eggs. Forcing them to lay year-round without rest can deplete their reserves faster, leading to earlier retirement or health issues.
- **Increased stress:** Winter is naturally a time for rest and recovery. Skipping this rest period puts extra strain on their bodies.
- **Molting complications:** Some hens won't molt properly if light prevents the hormonal shift, leading to poor feather quality.
- **No "natural rhythm":** Many chicken keepers prefer to let their flocks follow seasonal cycles, believing it leads to healthier, longer-lived birds.

If You Decide to Use Supplemental Lighting:

Here's how to do it responsibly:

1. **Aim for 14-16 hours of total light** (natural + artificial combined).
2. **Add light in the morning, not evening.** Set a timer to turn lights on before dawn rather than keeping them on after sunset. This allows hens to naturally roost at dusk, reducing stress.
3. **Use a low-wattage bulb** (40-60 watts) for a soft glow, nothing harsh or glaring.
4. **Don't start until after the first molt.** Let young hens complete their first adult molt naturally before adding supplemental light.
5. **Give them a break.** Even with lighting, consider giving your flock 4-6 weeks of rest per year (usually mid-winter) to recover.

Alternatives to Artificial Lighting

If you'd rather let nature take its course but still want to support winter laying, try these strategies:

1. Focus on Cold-Hardy, High-Production Breeds

Not all chickens are created equal when it comes to winter laying. If consistent eggs are important to you, choose breeds known for cold tolerance and year-round production:

- Rhode Island Reds
- Black Australorps
- Barred Plymouth Rocks
- Buff Orpingtons
- New Hampshire Reds

2. Optimize Nutrition

Winter laying requires extra energy. Support your hens by:

- Offering **layer feed with 16-18% protein**
- Providing **warm treats** in the morning (oatmeal, warm mash, scrambled eggs)
- Ensuring constant access to **oyster shell or crushed eggshells** for calcium
- Adding **scratch grains** in the evening to help them generate body heat overnight

3. Keep Them Warm (But Not Too Warm)

Healthy, well-fed chickens can handle cold temperatures surprisingly well, but extreme cold saps energy that could otherwise go toward egg production. Help them stay comfortable by:

- Insulating the coop (without blocking ventilation)
- Providing deep bedding (straw, pine shavings)
- Blocking drafts while maintaining airflow

- Using heated waterers so they stay hydrated
- Offering a windbreak in the run

4. Maximize Natural Light

Even without adding artificial light, you can help your hens make the most of winter daylight:

- Keep coop windows clean to let in maximum sunlight
- Use reflective materials (like white paint) inside the coop to brighten the space
- Let hens out into the run as early as possible on sunny days

5. Manage Expectations

This is the simplest (and often wisest) approach: accept that winter is a rest period. Use this time to:

- Stock up on store-bought eggs or trade with friends who use lighting
- Enjoy the break from daily egg collection and cleaning
- Focus on flock health, coop maintenance, and planning for spring
- Appreciate the eggs you *do* get as special winter treats

Real Talk: What I Do

After years of experimenting, I've landed in the "let them rest" camp. My girls slow down or stop laying from November through February, and honestly? I've made peace with it. They've worked hard all spring and summer, and I'd rather they live longer, healthier lives than squeeze out a few extra winter eggs.

Plus, when that first egg appears in late February, it feels like a gift, a promise that spring is coming and the world is waking up again.

When Winter Eggs Are a Problem

If your hens are young (under 2 years old), healthy, well-fed, and *still* not laying by late winter (February/March), it might indicate an issue:

- **Parasites** (mites, lice, worms draining energy)
- **Illness** (respiratory infections, reproductive issues)
- **Poor nutrition** (protein or calcium deficiency)
- **Stress** (overcrowding, predator pressure, bullying)
- **Broodiness** (some hens go broody in winter and stop laying)

Conduct a thorough health check and consult a vet if needed.

The Bottom Line

Winter egg production is a personal choice. There's no "right" answer, only what works best for you and your flock. Whether you light the coop or let your hens rest, the key is understanding *why* production drops and supporting your birds' health through the cold months.

And remember: spring always comes. When it does, those nesting boxes will be full again, and you'll have happy, healthy hens ready to reward your patience.

Winter Egg Production Quick Guide

Approach	Pros	Cons
Supplemental Lighting	Consistent eggs year-round	May shorten lifespan; prevents natural rest
Natural Cycle (No Lighting)	Healthier, longer-lived hens; follows natural rhythms	Fewer/no eggs in winter
Hybrid Approach	Moderate production; some rest	Requires careful management

Winter Laying Checklist

☐ Choose cold-hardy, productive breeds

☐ Provide 16-18% protein layer feed

☐ Offer warm treats and scratch grains

☐ Ensure constant access to calcium

☐ Keep waterers from freezing

☐ Insulate coop (but maintain ventilation)

☐ Maximize natural light in the coop

☐ Decide: supplement lighting or let them rest?

☐ Monitor health and weight regularly

☐ Adjust expectations and enjoy the season

Once you've covered the winter basics, the next step is understanding how chickens cope with weather changes in every season. Heat, cold, humidity, and wind all affect your flock differently, and a few simple adjustments can keep them thriving year-round.

All-Weather Chickens: Adapting to Your Climate

General Care	Chickens are hardy creatures, but they still need protection from temperature extremes. Make sure their coop offers warmth in winter and cooling options in summer.
Hot-Weather Tips	1. Provide plenty of shade and clean, fresh water. 2. Place frozen water bottles or ice blocks in the run for a quick cooldown. 3. Keep the coop well-ventilated, but ensure all openings are predator-proof.
Cold-Weather Care	1. Insulate the coop to trap heat, but maintain airflow to prevent moisture buildup. 2. Add extra dry bedding, such as straw or wood shavings, to help retain warmth. 3. Check combs and wattles for frostbite; apply a thin layer of petroleum jelly as protection. 4. Maintain coop temperatures between 10–30 °C (50–85 °F) — if you're comfortable, your hens likely are too.

With your flock prepared for all kinds of weather, the next step is making sure they're well-fed. Let's look at how you can provide nutritious, homemade feed.

Homemade Chicken Feed

Store-bought feed is convenient, but making your own gives you full control over nutrition, freshness, and cost. Plus, it's oddly satisfying like baking bread, but for chickens! Commercial feeds can contain fillers or ingredients you'd rather skip. Making your own lets you tailor the mix for your flock's age, breed, and egg-laying needs. Think of it as crafting a personalized diet plan for your girls! Here's how you can whip up a nutritious mix for your hens, keeping them healthy and happy.

Grab a bag and gather some grains and seeds like wheat, oats, barley, corn, and sunflower seeds. You'll also need protein boosters like fish meal, field peas, or mealworms, and finish with a pinch of kelp meal or vitamin powder to give your chickens a healthy dose of essential minerals.

Start with roughly 60% grains, 30% protein, and 10% extras (minerals, supplements, or treats). Adjust based on your hens' activity levels; more protein in winter, more greens in summer. Stir it all up in a big bin or bucket, seal it tight, and voilà, you have fresh feed with no mystery ingredients to serve your flock. Just make sure to keep your mix dry and cool. Moisture is the enemy. Nobody likes soggy chicken cereal! Here's a step-by-step guide to help you create a gourmet meal for your hens.

Step 1: Gather ingredients & tools

- 60% whole grains by weight (cracked corn, wheat, or barley)
- 20% protein source (soybean meal, sunflower meal, fish meal, or cooked/ground legumes)
- 10% calcium source for layers (crushed oyster shell or ground eggshells)

- 5% vegetable/seed mix (oats, millet, or sunflower seeds)
- 5% extras (kelp meal, vitamin/mineral premix, and grit if birds free-range less)
- A large mixing tub or food-safe barrel
- Kitchen scale
- Scoop
- Airtight storage bins
- Gloves
- Measuring cups
- Funnel
- Grinder (optional for crushing shells/meal).

For chicks/growers: increase protein to 20–24% (use a higher proportion of soybean meal or add cooked egg, mealworms, or fish meal) and omit the calcium boost until pullets approach laying age. For winter or hard-working birds: raise energy or fat slightly by adding 5–10% cracked corn or sunflower seeds.

Step 2: Weigh & measure (example: make 10 kg total)

- Weigh 6.0 kg whole grains.
- Weigh 2.0 kg protein source.
- Weigh 1.0 kg calcium source (only for layers).
- Weigh 0.5 kg veggie/seed mix.
- Weigh 0.5 kg extras (vitamin/mineral premix, kelp, grit). (If making feed for chicks, adjust protein to 2.2 – 2.4 kg and reduce calcium to 0.)

Step 3: Mix thoroughly

- Dump all the ingredients into your mixing tub.
- Using a clean shovel or large spoon, fold and turn the mix for several minutes until evenly blended. If you have a cement mixer or grain mixer, use it on low speed for 2–4 minutes.

- If adding wet bits (cooked egg, yogurt) for special treats, mix just before feeding. Do not store wet mixes.

Step 4: Condition & supplement

- Add a poultry vitamin/mineral premix per label directions (especially important if you're replacing a commercial complete feed).
- Add probiotics (optional) by sprinkling according to product instructions.
- For layers: ensure free-choice crushed oyster shells available separately. Don't mix large amounts of calcium into the complete feed for all birds unless you want every bird to get extra calcium.

Step 5. Store safely

- Place feed in airtight, rodent-proof containers or barrels.
- Label with date made. Use within 6–12 weeks for best freshness.
- Keep in a cool, dry spot out of direct sun and away from moisture. Rotate stock (first made = first used).

Adult laying hens typically need around 120 to 150 grams of feed per day, though this amount can vary depending on their activity level and how much they forage on their own. Growers (young chickens that aren't yet laying) should be given free access to feed with the right protein balance, and their weight should be monitored regularly to ensure healthy growth. Free-range birds often eat less of the feed you provide in a bowl since they find natural food sources outdoors, but they still need access to grit and minerals to aid digestion. Keep an eye on your flock's eggshell quality, feather condition, and body weight; thin shells often mean your hens need more calcium, while weight loss or poor feathering can signal the need for extra protein or calories.

When it comes to safety, there are a few hard rules: never include raw or green potatoes, avocado, chocolate, caffeine, or moldy grains in your feed mix, as these can be toxic to chickens. Avoid giving too much calcium to birds that aren't laying yet, since excess calcium can cause kidney and liver problems.

If your vet recommends a medicated feed, always follow their advice instead of using homemade mixes. And when in doubt, especially for larger flocks or breeding birds, it's always smart to consult a poultry nutritionist or veterinarian to make sure your birds are getting exactly what they need to stay healthy and productive.

DIY Egg Storage Solution

You've got a large batch of glorious, freshly laid eggs and need a way to store them safely. Forget boring cartons. Let's build your own rustic DIY Egg Storage Rack. So let's roll up those sleeves and get cracking.

Step 1: Gather Your Supplies

You'll need a wooden crate, small basket, or even a recycled spice rack. Basically, anything with compartments works. You'll also need some hay or shredded paper for cushioning, and if you want to go full farmhouse chic, grab a bit of chalk paint for decoration.

Step 2: Clean and Prep

Wipe down your chosen container and give it a light sanding if it's wood. Paint or stain it if you like. This is your chance to match your coop's aesthetic (or your kitchen's).

Step 3: Add Cushioning

Line each compartment with hay, straw, or soft paper shreds. This prevents eggs from cracking and adds that cozy "farm fresh" vibe.

Step 4: Organize Your Eggs

Place unwashed eggs pointy side down. This keeps the air pocket inside stable and extends freshness. If you collect eggs daily, store them in order from oldest to newest, so you always use the older ones first.

Step 5: Label and Display

Add a chalkboard tag or marker strip to jot down the collection dates. Not only does this look adorable, but it's also practical for tracking freshness.

The Takeaway

Caring for chickens means learning the rhythm of their lives. Daily egg collection, smart storage techniques, and keeping your flock happy through blazing summers and frosty winters are the keys to having healthy chickens. While these practices are essential for your birds' well-being, even the healthiest flocks can sometimes hit a rough patch. A droopy comb, a sudden dip in egg production, or unusual behavior might leave you wondering, "What's going on?" Don't worry, every chicken keeper faces these moments.

The next chapter is all about understanding common chicken health issues from early warning signs to simple care routines that keep your flock strong and disease-free.

CHAPTER 5

Healthy Feathers, Naturally

The secret to a thriving flock lies in the art of observation, and catching small changes before they turn into big problems. A droopy comb, a ruffled feather, or a hen that's suddenly not clucking with the morning crowd can tell you more than you'd think. Learning to read the subtle signs can make all the difference between a thriving flock and a coop full of worries.

In this chapter, we'll learn how to spot early signs of illness and tackle common chicken ailments naturally. We'll explore the practical (and natural) ways to keep your birds in peak condition, spot trouble early, and respond with confidence.

Wing-Spotting: Telling the Healthy from the Hurting

If you spend enough time around your flock, you'll quickly realize that chickens have personalities as colorful as their feathers. Some strut like royalty, others gossip in clucks, and a few seem to take pride in stealing your seat the second you turn your back. But among all this charm and chaos, a keen eye can tell when something isn't quite right. Observing your chickens daily is the secret to keeping them happy, healthy, and clucking for years to come.

A healthy chicken is alert, bright-eyed, and full of curiosity. Her feathers gleam, her comb and wattles are plump and vibrant red, and she's quick to scratch the ground in search of treats. Her body should feel well-rounded and not bony or bloated. Her abdomen should be firm but not hard. If you gently run your hand along her breastbone and it feels sharp or jutting, she may be underweight. On the other hand, a swollen or squishy belly could hint at internal issues like fluid buildup or egg peritonitis[23].

Keep an eye on the finer details too: clear eyes, clean nostrils, and smooth, shiny beaks are all signs of good health. A dull, crusty comb or discharge around the nares (those tiny nostril openings) might signal infection. Healthy droppings are surprisingly informative. They should be firm, brown, and topped with a white cap of urates. Anything runny, discolored, or bloody means a doctor's appointment is due.

Their feathers should lie sleek and full, not ragged or patchy. Legs and feet must be free of swelling, scabs, or flaky buildup that could indicate mites or infection. And make sure to check their backends, too. A healthy vent is moist and clean, while a messy or swollen one could spell trouble.

Every chicken keeper becomes a bit of a "wing-spotter" over time, learning the subtle art of reading their flock's body language, feathers, and droppings like clues in a mystery novel. The earlier you catch a problem, the easier it is to fix.

Know What a Healthy Chicken Looks Like

A truly healthy chicken is alert, curious, and struts about like she owns the yard (because, let's be honest, she does). Her feathers shine in the sunlight, her comb and wattles are plump and a vibrant red, and her eyes are bright and clear. She'll scratch the ground with determination, dust-bathe like it's a day at the spa, and chat away with soft, contented clucks. Her droppings are firm with a white cap, her breathing is quiet, and she maintains a steady appetite and weight. Once you know your birds' normal quirks and rhythms, even the smallest change becomes easy to spot.

Common Sick Chicken Symptoms

Even the most pampered flock can fall ill from time to time, and recognizing early signs can make all the difference. The first clue often comes from their comb and wattles. A once-bright red comb turning pale, bluish, or spotty can indicate anemia, stress, or illness. Another telltale sign is a sudden drop in egg production. If hens that once filled your basket daily slow down suddenly, then they could be under the weather.[24]

Weight loss and loss of appetite are other warning flags. If a chicken feels unusually bony or uninterested in food, she might be fighting worms, infection, or nutritional deficiencies. A distended or mushy crop (that pouch at her throat) could mean digestive issues or impaction.

Behavioral changes are equally revealing. A healthy hen is active and alert; one that's lethargic, standing still with feathers puffed, or hiding in a corner may be in distress. Watch for irregular or cloudy eyes, crusty nostrils, or sneezing, which can all indicate respiratory trouble. Scratching and excessive preening may signal mites or lice, while lameness, waddling, or scaly buildup on the legs could be signs of joint problems or scaly leg mites.

Pay attention to their droppings, too: watery, green, or bloody feces can point to coccidiosis or other infections. A clogged or swollen vent, soft-shelled or misshapen eggs, and bald spots are all red flags. Difficulty breathing, wheezing, or a gaping beak when at rest signals respiratory distress, while spots, lesions, or paralysis are serious symptoms requiring immediate attention.

Sick Chicken Treatment Options & Prevention

When illness strikes, it's time to become a caring chicken nurse. The first step is isolation. Move your sick hen to a calm, draft-free space away from the flock to prevent contagion and reduce stress. Make sure she has soft bedding, clean air, and easy access to water.

Hydration is vital. Sick chickens can dehydrate quickly, so keep water fresh and accessible, adding electrolytes or a little apple cider vinegar can encourage drinking. Offer her favorite snacks to tempt her appetite, but keep food soft and easy to digest.

Let her rest. Much like us, chickens heal best when they feel safe and unbothered. Monitor her daily. Observe her droppings, her energy, and whether her appetite improves. If symptoms worsen, or if she struggles to breathe or stand, it's time to call a vet or an experienced poultry keeper for guidance.

Prevention, however, is your strongest tool. Keep coops clean and dry, replace bedding regularly, and provide balanced nutrition. Conduct quick daily check-ins and continue observing their comb color, droppings, and posture. And remember, isolation doesn't always mean complete loneliness. Some chickens get anxious alone, so letting a calm flockmate stay nearby (separated by mesh) can keep their spirits up. Just be cautious not to risk spreading disease.

In the end, being a backyard chicken keeper means becoming part farmer, part detective, and part nurse. You'll celebrate every soft cluck of recovery and learn to trust your instincts when something feels off. Up next, we'll look into treating common chicken illnesses because when your flock's health is on the line, knowing what to do next makes all the difference.

Signs of Parasites (And What to Do About Them)

Even the cleanest coops can fall victim to these sneaky freeloaders. Worms, mites, and lice are the uninvited guests of the poultry world. They can seriously affect your flock's health, egg production, and happiness. The good news? Once you know what to look for, you can catch infestations early and send those pesky hitchhikers packing.

External Chicken Parasites: Mites and Lice

If your hens are looking a bit scruffy, constantly preening, or dust-bathing more than usual, it's time to check for mites and lice.

These tiny troublemakers live on or near the skin, feeding on feathers, blood, or skin debris.[25]

The Northern Fowl Mite loves to set up camp around the vent and under the wings. You might spot dark specks near the base of feathers or notice irritation and feather loss in those areas. Red Mites, on the other hand, are nocturnal vampires. They feed on your chickens at night and hide in coop crevices during the day. If your flock refuses to roost at night or seems restless, suspect red mites.[25]

Then there's the Scaly Leg Mite, a nasty little creature that burrows under the leg scales, causing them to lift and appear crusty or deformed. Left untreated, this can lead to pain and lameness. Lice, especially the Shaft Louse, are easier to spot. Look for tiny yellowish insects and egg clusters clinging to feather shafts near the vent and thighs.[25]

To combat external parasites, start with a thorough coop clean up. Replace all bedding, scrub roosts and nesting boxes with a poultry-safe disinfectant, and dust the area with diatomaceous earth or a mite powder. Treat each chicken with a vet-approved mite and lice treatment and repeat as directed. Remember prevention is key, so keep the coops dry, avoid overcrowding, and provide regular dust baths so hens can self-clean.

Internal Chicken Parasites: The Hidden Invaders

Internal parasites like worms and protozoa are harder to spot, but their effects can be just as harmful. A worm-infested chicken might lose weight, lay fewer eggs, develop a pale comb, or pass odd-looking droppings. Common culprits include:

- **Roundworms (Ascaridia galli):** The most common intestinal worm, causing diarrhea, lethargy, and poor growth.
- **Threadworms and Cecal Worms:** Often cause digestive upset and reduced nutrient absorption. Cecal worms can also carry blackhead disease, dangerous to turkeys and chickens alike.

- **Gapeworms (Syngamus trachea):** These live in the throat, making birds gasp or stretch their necks in an attempt to breathe.

- **Tapeworms (Cestodes):** Flat, segmented worms that rob your birds of nutrients.

- **Protozoa (Coccidia):** Microscopic parasites that attack the intestinal lining, leading to coccidiosis: a serious, sometimes fatal condition causing bloody droppings and weakness.

Treating worms and coccidiosis usually requires medication such as a poultry dewormer or coccidiostat. You can easily find these at your local farm supply stores or through your vet. Always treat the entire flock, as parasites spread easily.

Prevention: The Real Secret Weapon

Preventing parasites is far easier (and kinder) than treating them. Keep your coop clean and dry, rotate dust baths, and regularly inspect your flock, especially around the vent, under wings, and legs. Replace bedding often, and periodically disinfect perches and corners where mites hide. For internal parasites, practice good pasture management: avoid letting chickens forage in the same area for too long, and consider a deworming schedule two to four times a year.

Healthy, well-fed chickens with strong immune systems are far less likely to suffer infestations. A balanced diet, access to grit, and probiotic supplements all help build resistance. After all, your feathered friends deserve to spend their days happily scratching, dust-bathing, and gossiping in the sun, not battling invisible freeloaders. With a bit of vigilance and regular care, you'll keep your flock parasite-free and thriving.

Naturally Preventing and Treating Lice and Mites

You don't always have to resort to chemical treatments to get rid of those pesky mites and lice. While these treatments are effective, natural methods offer a much gentler approach that's safer for both birds and humans.

1. Dust Bathing Area

Chickens are natural self-care enthusiasts. A good dust bath is their version of a spa day and their best defense against parasites. Provide a dry, sheltered area filled with loose soil or sand where your hens can roll, fluff, and shake to their heart's content. This helps dislodge dirt, oils, and, most importantly, unwanted pests.

2. Add Wood Ash

Wood ash from untreated wood is like a secret weapon in your parasite-prevention toolkit. Mixed into the dust bath, it smothers mites and lice naturally by drying them out. Combine equal parts fine sand, wood ash, and soil for a perfect parasite-busting blend.

1. Boost Nutrition with Yeast

Adding brewer's yeast or cultured dried yeast to your flock's feed can strengthen their immune systems and improve feather health. Healthier birds are less likely to become parasite hosts in the first place!

Natural Mite Treatment

Mites can be a real menace for your birds. Here's how you can keep them at bay and avoid coop infestations.

Step 1: Clean the Coop Thoroughly

Start by removing all bedding and scrubbing down every corner with a mild soap or vinegar solution. Mites love cracks, corners, and roost bars, so don't skip those spots! Let the coop dry completely before refilling it with fresh bedding.

Step 2: Treat the Coop

Once clean, sprinkle diatomaceous earth (DE) or wood ash around roosts, nesting boxes, and floor edges. DE is a fine powder made from fossilized algae that kills mites by dehydrating them.

Step 3: Dust the Chickens with Wood Ash

Gently rub a bit of wood ash or DE into the feathers under their wings and around the vent, two of the mites' favorite hideouts. This helps suffocate and dehydrate the pests naturally.

Step 4: Treat the Chickens (Topically or Internally)

For scaly leg mites, apply a natural oil such as coconut or olive oil directly onto the legs every few days. The oil suffocates the mites and softens the lifted scales so healing can begin. Adding crushed garlic or a few drops of apple cider vinegar to your chickens' drinking water can also help boost their immune response and repel external parasites.

DIY Chicken Mite Treatment Spray

Here's how you can make a mite treatment solution for your birds. The items you'll need include

- 1 cup water
- 1 cup apple cider vinegar
- 5 drops lavender essential oil
- 5 drops tea tree essential oil
- 3 drops clove oil (optional but powerful)

Instructions:

1. Mix everything in a spray bottle and shake well before each use.
2. Lightly mist your chickens (avoid the eyes and nostrils) and their coop surfaces once or twice a week.
3. Store in a cool, dark place and remake every few weeks for freshness.

Lavender and tea tree oil have natural antibacterial and insecticidal properties, while vinegar repels mites and deodorizes the coop, leaving your flock smelling farm-fresh and mite-free. If you notice raised, crusty leg scales, your bird likely has scaly leg mites. Along with oil treatments, gently cleaning the legs with warm, soapy water before applying coconut oil can help loosen the old scales and suffocate the mites faster.

For coccidiosis, a microscopic intestinal parasite, prevention is key: keep bedding dry, rotate runs to prevent damp, contaminated soil, and add probiotics or garlic to water for natural gut support.

Remember good hygiene and a consistent routine is essential to prevent the spread of disease. Regular cleaning, healthy diets, and giving your chickens space to dust bathe, scratch, and explore will help keep them parasite-free. A healthy gut is the foundation of a strong immune system and that starts with what your flock eats. Fresh greens, grains, bugs, and high-quality feed keep their digestive systems thriving, while herbs like oregano, thyme, garlic, and turmeric add an extra immune-boosting punch with their natural antimicrobial powers.

For an extra wellness kick, you can mix a spoonful of apple cider vinegar (ACV) into each gallon of drinking water a few times a week. It supports digestion, fights harmful bacteria, and even helps with calcium absorption for sturdier eggshells. You can also sprinkle in brewer's yeast, kelp meal, or probiotic powders to keep their gut flora happy and balanced. To keep your backyard flock healthy, focus on prevention, good nutrition, and gentle, natural remedies. Chickens are surprisingly resilient creatures, but just like us, they can fall ill when their environment, diet, or stress levels go out of balance. Happy hens make healthy eggs, and healthy eggs start with clean, cared-for chickens. With a bit of ash, oil, and old-fashioned elbow grease, you'll have your flock strutting around mite-free in no time.

Common Chicken Diseases and Their Treatment

Let's take a look at some of the illnesses that can affect your flock and how you can treat them.

Disease: Respiratory Illnesses (Mycoplasma and Infectious Coryza)

Infective Agent / Cause:

Mycoplasma gallisepticum (bacteria) and Avibacterium paragallinarum (bacteria)

Symptoms:

Sneezing, nasal discharge, wheezing, swollen eyes

Treatment/ Response:

- Add oregano oil or garlic extract to drinking water (natural antibiotic)
- Bring birds into a warm, steamy bathroom to loosen mucus
- Diffuse eucalyptus or tea tree oil near (not inside) the coop
- Keep coop clean, dry, and well-ventilated

Disease: Sour Crop

Infective Agent / Cause:

Yeast/fungal overgrowth (Candida albicans) from fermented food stuck in crop

Symptoms:

Squishy, sour-smelling lump under the neck

Treatment/ Response:

- Remove food for 12 hours, offer water with a few drops of ACV
- Massage the crop gently downward
- Offer soft foods like scrambled eggs or plain yogurt once cleared

Disease: Bumblefoot

Infective Agent / Cause:

Staphylococcus bacterial infection through cuts or splinters

Symptoms:

Swelling, scabbing, limping

Treatment/ Response:

- Soak foot in warm Epsom salt water 10–15 mins daily

- Apply manuka honey or antibacterial salve, wrap lightly
- Keep perches smooth, bedding dry

Disease: Marek's Disease

Infective Agent / Cause:

Herpes virus (viral infection)

Symptoms:

Paralysis, weakness, loss of coordination

Treatment/ Response:

- No cure — focus on vaccination and hygiene
- Oregano and garlic supplements for immune support
- Quarantine new birds before mixing

Disease: Avian Influenza (Bird Flu)

Infective Agent / Cause:

Influenza Type A virus

Symptoms:

Sudden death, respiratory distress, drop in egg production

Treatment/ Response:

- Limit wild bird contact, cover runs
- Disinfect feeders, change shoes before entering coop
- Maintain strict biosecurity and report outbreaks

Disease: Virulent Newcastle Disease (vND)

Infective Agent / Cause:

Paramyxovirus

Symptoms:

Twisted necks, paralysis, respiratory distress, reduced laying

Treatment/ Response:

- Get sick chickens examined by the vet immediately
- Keep immune system strong with good nutrition
- Disinfect coop and gear frequently

Disease: Salmonella

Infective Agent / Cause:

Salmonella enterica (bacteria)

Symptoms:

Often symptomless in birds; risk of human infection

Treatment/ Response:

- Clean eggs gently (don't soak)
- Wash hands after handling
- Test flock for Salmonella Enteritidis (SE) if selling eggs/chicks

Disease: Flystrike

Infective Agent / Cause:

Fly larvae (maggots) infestation from unclean feathers or wounds

Symptoms:

Wounds with maggots, irritation, lethargy

Treatment/ Response:

- Bathe bird in warm, soapy water
- Remove maggots with tweezers
- Apply raw honey or coconut oil to soothe
- Prevent with clean coop and dry bedding

Disease: Feather Loss (Molting or Stress)

Infective Agent / Cause:

Natural molting, parasites, or nutritional deficiency

Symptoms:

Bald spots, pecking, dull feathers

Treatment/ Response:

- Provide extra protein (mealworms, eggs, sunflower seeds)
- Ensure adequate space
- Use chamomile or lavender to calm flock

Most minor illnesses resolve on their own or respond well to natural remedies, but if symptoms worsen like persistent breathing trouble, paralysis, or severe wounds, then it's time to consult a poultry vet. Catching issues early is your best defense.

After covering the major health concerns, let's turn to something that may *look* like a problem but usually isn't: molting. This natural reset can leave your hens looking rough, but it's an important part of their yearly rhythm.

Molting

One autumn morning, I walked out to the coop and stopped dead in my tracks. The ground was covered in feathers, so many feathers it looked like a pillow fight had broken out overnight. My heart raced. Had a predator gotten in? Were my hens sick? Then I spotted Dorothy strutting around, half-naked and surprisingly unbothered, pecking at her breakfast like nothing was wrong.

That was my introduction to molting, and let me tell you, nobody warns you how dramatic it looks the first time.

Molting is the natural process where chickens shed old, worn-out feathers and grow fresh new ones. It's like hitting the reset button on their plumage. While it may look alarming, it's completely normal and essential for their health. Understanding molting helps you support your flock through this vulnerable time without panicking every time you see a pile of feathers.

Why Do Chickens Molt?

Feathers take a beating throughout the year: sun damage, pecking, dust baths, and general wear and tear all contribute to deterioration. Molting allows chickens to replace damaged feathers with strong, insulating new ones just in time for winter. It's nature's way of preparing them for the cold months ahead.

Most adult chickens molt once a year, typically in late summer or early fall when daylight hours begin to shorten. The decreasing light triggers hormonal changes that halt egg production and redirect the bird's energy toward feather regrowth.

What to Expect During Molting

Molting can last anywhere from 8 to 16 weeks, depending on the individual bird and her overall health. Here's what you'll notice:

- **Feather Loss:** It can be gradual or sudden. Some hens lose feathers slowly over weeks, while others seem to shed entire sections overnight. The head, neck, back, and wings are usually affected first.

- **Scruffy Appearance:** Your once-glamorous hens may look downright ragged. Bare patches, stubby pin feathers (new growth covered in a waxy sheath), and an overall disheveled look are all normal.

- **Reduced (or Stopped) Egg Production:** Feathers are made of protein, about 80-85% of a feather is pure protein. Producing eggs also requires significant protein. A hen's body can't do both at once, so egg-laying slows dramatically or stops entirely during molting.

- **Behavioral Changes:** Molting hens may become quieter, less active, or even a bit irritable. They might spend more time resting and less time socializing. Their combs and wattles may also pale as their bodies focus energy inward.

- **Sensitivity:** New pin feathers are extremely sensitive because they contain blood vessels while growing. Your usually friendly hen might flinch or peck at you if touched during this time. Be gentle and avoid handling her unless necessary.

Supporting Your Flock Through Molting

Molting is physically demanding. Here's how to help your hens get through it smoothly:

1. Boost Protein Intake

During molting, your flock needs extra protein to rebuild feathers quickly and efficiently. Regular layer feed (around 16% protein) isn't enough. Increase protein to 18-20% by:

- Switching temporarily to a **grower feed** or **game bird feed** (higher protein content)
- Supplementing with **high-protein treats** like:
    - Mealworms (dried or live)
    - Scrambled eggs (yes, they can eat their own eggs!)
    - Black soldier fly larvae
    - Sunflower seeds

- Cooked fish or shrimp
- Lentils or split peas (cooked)

Avoid overloading them with scratch grains or corn during molting, these are low in protein and won't support feather growth.

2. Minimize Stress

Stress slows molting and feather regrowth. Keep the environment calm by:

- Avoiding flock changes (don't introduce new birds during molting)
- Limiting handling
- Maintaining a consistent routine
- Ensuring plenty of space so molting hens can retreat if needed

3. Provide Plenty of Clean Water

Hydration is essential for healthy feather development. Make sure waterers are always full and clean.

4. Reduce Egg Expectations

Accept that egg production will drop or stop entirely. Trying to force laying by adding supplemental light or high-calcium feeds can actually slow molting and stress your hens. Let their bodies rest.

5. Watch for Health Issues

While molting is natural, extreme feather loss, prolonged molting (beyond 16 weeks), or signs of illness (lethargy, weight loss, discolored skin) may indicate problems like mites, lice, malnutrition, or disease. Inspect your birds regularly and consult a vet if something seems off.

The Stages of Molting

Not all hens molt the same way, but here's a general timeline:

- **Weeks 1-3:** Feather loss begins, usually starting with the head and neck. Egg production drops.

- **Weeks 4-8:** Heavy shedding continues. Pin feathers start emerging. Hens look their scruffiest.
- **Weeks 9-12:** New feathers grow in rapidly. The waxy sheaths break off, revealing glossy new plumage.
- **Weeks 13-16:** Molting completes. Hens regain their full, beautiful feathers. Combs and wattles redden. Egg production gradually resumes.

Hard Molt vs. Soft Molt

- **Hard Molt:** Feathers drop quickly, and the hen looks nearly bald in spots. These hens usually finish molting faster and resume laying sooner.
- **Soft Molt:** Feathers drop gradually over a longer period. The hen maintains a more normal appearance but takes longer to complete the process.

What About Young Chickens?

Chicks molt several times during their first year as they transition from fluff to juvenile feathers to adult plumage. These early molts are less dramatic and happen quickly. The first **annual adult molt** typically occurs after a hen's first laying season, around 15-18 months of age.

When Molting Isn't Normal

Watch for these warning signs:

- **Excessive bald patches with red, irritated skin** (could be mites or bullying)
- **Molting outside of fall** (possible illness or extreme stress)
- **Prolonged molting beyond 16 weeks** (nutritional deficiency or health issue)
- **Feather loss with no new growth** (protein deficiency, disease, or parasites)

If you notice any of these issues, consult a poultry vet or experienced chicken keeper for guidance.

The Bright Side of Molting

Yes, molting is messy, and yes, your egg basket will be lighter for a while. But here's the silver lining: when your hens emerge with their fresh, vibrant feathers, they're healthier, stronger, and ready to face winter. Their new plumage provides better insulation, and once laying resumes, you'll often notice an improvement in egg quality.

Molting is nature's way of giving your flock a fresh start. With a little extra protein, patience, and TLC, your hens will sail through it and come out the other side looking (and feeling) fabulous.

Quick Molting Checklist

- ☐ Increase protein to 18-20%
- ☐ Offer high-protein treats daily
- ☐ Keep waterers clean and full
- ☐ Minimize stress and handling
- ☐ Expect reduced or stopped egg production
- ☐ Watch for signs of illness or parasites
- ☐ Be patient, molting takes 8-16 weeks
- ☐ Celebrate when those beautiful new feathers come in!

Helping your flock through their molt is one part of good care. Being prepared for scrapes, injuries, or sudden illnesses is the next, so let's talk about what to keep in your chicken first-aid kit.

The Ultimate Chicken Growers First Aid Kit

One minute they're pecking happily, the next they're limping, sneezing, or sporting a mystery scratch; raising these mischievous birds is a journey full of surprises. With a well-stocked first aid kit and a few trusty herbs on your side, you'll be ready for anything from a scraped comb to a sneezing hen.

When it comes to first aid, preparation is everything. Always have a clean, quiet place ready to isolate a sick or injured bird. Your chicken first aid kit should include an antiseptic spray (like VetRx or iodine), wound ointment, Epsom salts, gauze, scissors, tweezers, and electrolytes. For internal care, keep probiotics, apple cider vinegar, and vitamin supplements handy for recovery boosts. A small flashlight, gloves, and a towel are also must-haves for gentle handling.

Herbs are nature's pharmacy for your flock. They're gentle, effective, and packed with immune-boosting power. Whether sprinkled over feed, steeped into teas, or grown near the coop, they'll boost your flock's health, repel pests, and even calm nervous birds. Using herbs regularly can prevent many common issues before they start, saving you time and heartache later on.

- Astragalus (Astragalus membranaceus) is a powerhouse root known for strengthening the immune system and helping chickens fight off respiratory infections. It's best added to feed or brewed into a tea during times of stress or seasonal change.

- Calendula (Calendula officinalis), with its sunny orange petals, promotes skin healing and bright yolks. This herb is great for minor wounds and boosting digestive health.

- Echinacea acts like an immune tonic, useful during illness or after exposure to new birds.

- Oregano is one of the most potent natural antibiotics available, reducing harmful bacteria and keeping the gut balanced.

- Thyme and rosemary are strong antimicrobials, while parsley is rich in vitamins A and K for feather growth and egg health.

- Stinging nettle, once dried or cooked, boosts circulation and provides essential minerals.

- For relaxation, lavender, sage, mint, and cilantro calm anxious hens and freshen nesting areas naturally.

Some plants to avoid include foxglove, rhubarb leaves, hemlock, and nightshade, they're toxic to chickens.

Monthly Chicken Health Check-Up Checklist

(Because a happy flock = a happy keeper!)

Date: _____ **Checked by:** _____

Step 1: General Observation

Before handling, watch your chickens for 5–10 minutes.
- ☐ All chickens active, alert, and social
- ☐ Normal appetite and drinking behavior
- ☐ No limping, isolation, or puffed-up posture
- ☐ No sneezing, coughing, or wheezing sounds
- ☐ Clean, dry environment with no strong ammonia smell

Step 2: Hands-On Chicken Check

Body Condition
- ☐ Breastbone feels firm (not too sharp or too rounded)
- ☐ No visible swelling or bloating

Comb & Wattles
- ☐ Bright red, smooth, and warm to the touch
- ☐ No pale, dark, or shriveled areas

Eyes, Ears & Beak
- ☐ Eyes clear and bright (no discharge or swelling)
- ☐ Ears clean and dry
- ☐ Beak properly aligned and free from cracks

Feathers & Skin
- ☐ Feathers smooth and glossy
- ☐ No bald spots, redness, or scabs
- ☐ Check near vent for mites or lice (tiny specks or movement)

Legs & Feet
- ☐ Scales flat and smooth (no crustiness—sign of scaly leg mites)
- ☐ No swelling, cuts, or dark scabs under feet (bumblefoot)

Abdomen & Vent

☐ Abdomen soft but not distended
☐ Vent clean, moist, and free from buildup or redness

Step 3: Droppings & Eggs

☐ Excrement firm and brown with a white cap
☐ No green, bloody, or watery droppings
☐ Eggs solid, normal-shaped, and good shell quality

Step 4: Coop & Environment

☐ Coop clean, dry, and odor-free
☐ Bedding replaced or refreshed
☐ Feeders and waterers scrubbed and refilled
☐ Adequate ventilation and shade available
☐ Dust bath area clean and replenished (add wood ash or sand if needed)

Step 5: Record & Reward

☐ Health notes recorded (weight, symptoms, changes)
☐ Any concerns noted for follow-up
☐ Chickens rewarded with a healthy snack (mealworms, veggie scraps, or herbs!)

Notes:

_____
_____
_____

For easy reference, here's a quick guide to must-have supplies:

Item	Purpose	Notes
Electrolyte powder	Hydration support	Mix with water during heat or illness
Blu-Kote spray	Antiseptic & wound cover	Stops pecking at injuries
Epsom salt	Detox & soak	Great for swollen feet or legs
VetRx	Respiratory aid	Use for sniffles or mild colds
Tweezers & scissors	Handling feathers or splinters	Keep sterilized
Disposable gloves	Hygiene	Always wear during treatment

Keep everything in a sealed box near your coop, it's better to have it and not need it than the other way around.

Trouble in the Coop? Quick Fix Guide

Problem	Likely Cause	Quick Fix
No eggs	Molting, low daylight, poor diet	Boost calcium and protein; add light hours
Feather loss	Mites, bullying, or molt	Check for parasites; provide dust bath
Pale comb	Anemia or worms	Deworm; add iron-rich feed
Smelly coop	Wet bedding	Clean out, add dry litter
Broody hen won't move	Hormones	Cool nest box, limit nesting time
Soft shells	Low calcium	Provide oyster shell or crushed eggshells
Lethargic hen	Heat or illness	Check hydration and isolate if needed

Quick reference = calm chicken keeper. I've put together a comprehensive quick guide list for you at the end of the book, check it out on page XX. It's a practical, easy-to-use resource designed to help you feel prepared and confident in nearly every situation you'll face on your hen-keeping journey.

Bookmark this page for fast peace of mind!

The Takeaway

Clean coops, good feed, and a touch of nature's medicine is all it takes to maintain a happy flock. Keep an eye out for signs of trouble so you can nip problems in the bud. Symptoms like dull feathers or lethargy could point toward underlying problems. Quick action with herbs like oregano, astragalus, and calendula boost your chickens' immunity and fight off disease-causing agents.

Regular coop cleaning, dust baths, and a stocked first aid kit will help you prevent and respond to accidents and keep those pesky mites and lice at bay. Ultimately, healthy feathers and bright eyes reflect more than just good health, they show your dedication. A mindful keeper is always prepared for when things go South.

In the next chapter, we'll look at the terrifying and exciting stage of a chicken keeper's journey: caring for those cute little chicks. From incubation and hatching to the tender early weeks of chick care, we'll take a close look at all the important aspects of this crucial stage.

CHAPTER 6

Raising Chicks from Hatch to Home

The box was smaller than I'd expected. When I opened it, I was met with an onslaught of cuteness! Five fuzzy chicks blinked up at me, all legs and fluff, and for a second, I just stood there grinning like an idiot. Then the panic hit. The brooder wasn't ready. The heat lamp flickered like a dying campfire, and I couldn't tell if the thermometer was lying or if my new babies were already freezing. One chick climbed the water dish and soaked her belly. Another started pecking at the wrong thing entirely. My cat watched from across the room like it was a reality show called Dinner with Drama.

That first night, I barely slept. I kept tiptoeing back to check if they were breathing, terrified I'd done something wrong; the wrong feed, the wrong bedding, too much heat, not enough air. But somehow, they survived my bumbling efforts, and by morning, they were dry, fluffy, and cheeping like they owned the place. I realized then that raising chicks wasn't about getting everything perfect, it was about learning to care, adjust, and laugh through the chaos. This chapter will guide you through everything I wish I'd known that first day.

Preparation & Supplies

Before your chicks arrive, decide where you plan to keep them and gather the essentials. Many people buy from large commercial hatcheries that ship day-old chicks nationwide and often offer female-only options with vaccinations. You can also choose from the seasonal selection at your local feed store, though they usually offer fewer breeds and can't guarantee gender. Supporting small farms or local breeders lets you see the birds in person, check their health and temperament, and often reduces shipping stress.

You'll need a brooder box, a heat source (lamp or plate), a feeder, waterer, bedding such as pine shavings, chick starter feed, grit, and a thermometer. It's also helpful to have first-aid basics like probiotics and electrolytes on hand. Having everything ready before the chicks come home gives them a calm, healthy start.

How to Set Up a Chick Brooder

The brooder is the warm, secure space where baby chicks live until they're feathered and strong enough for outdoor temperatures. Choose a draft-free, quiet spot such as a garage, spare room, or enclosed porch.

For the brooder itself, use anything roomy and easy to clean, such as a large plastic bin, cardboard box, wooden crate, stock tank, storage tub, or kiddie pool. Allow at least ½ square foot per chick for the first few weeks, expanding the space as they grow. Add pine shavings for bedding and set up the feeder, waterer, heat source, and thermometer.

- **Heat & Temperature:** A brooder's warmth could mean life or death for chicks. Use a heat lamp or plate, keeping the temperature around 95°F (35°C) for week one, dropping by 5°F weekly until they're fully feathered. Always let chicks choose their comfort, if they huddle under the lamp, it's too cold; if they avoid it, too hot.

- **Food and Water:** Provide chick starter feed and fresh water in shallow dishes to prevent drowning accidents. You can supplement the feed with medication if you want extra coccidiosis protection. By 16–18 weeks, transition to layer feed and calcium sources such as crushed oyster shells or eggshells (see Chapter 4).

- **Cleaning & Care:** Replace bedding every few days and keep waterers spotless. Dirty brooders invite disease fast. Chicks typically stay in the brooder for 4–6 weeks, until they're feathered and strong enough to join the outdoor coop.

Chick Treats and Grit

Just like kids love snacks, chicks adore treats. For the first few weeks, stick to chick starter feed only. Once they're about 3 weeks old, you can introduce small treats like finely chopped greens, grated carrots, mashed boiled eggs, or oatmeal. Whenever you give treats, always offer grit (tiny bits of sand or fine gravel) to help them digest their food properly.

Adult hens can enjoy a wider buffet including fruits, veggies, grains, and protein-rich snacks like mealworms, but always keep in mind that treats should make up no more than 10% of their diet. Avoid feeding onions, chocolate, avocado, or raw potato peels, which can be toxic.

During molting, offer extra protein; in winter, warm mash or oats help maintain body heat; and in summer, cool watermelon or cucumber keeps them hydrated. Healthy snacks, balanced feed, and plenty of grit will keep those tiny beaks and bellies happy and your chicks well-nourished.

Handling Chicks

Gentle handling is the key to raising friendly, confident chickens. Spend a few minutes each day letting your chicks get used to your hands. Move slowly, speak softly, and scoop them up gently with both hands. Never grab them from above, as it mimics a predator attack and can cause panic. Supervise children closely when handling chicks, and always wash hands afterward to prevent the spread of germs like Salmonella.

Chick Health, Problems and Diseases

Chicks are delicate and can fall ill quickly. Observing them daily helps you spot trouble before it spreads. Healthy chicks are bright-eyed, alert, and active; a sick chick may be lethargic, huddled alone, or have ruffled feathers. Here are some common problems your feathered babies might face during those crucial first few weeks.[27]

Coccidiosis

One of the most common and dangerous chick diseases, coccidiosis is caused by a microscopic parasite that attacks the intestines. It spreads easily through droppings, damp bedding, or contaminated feed and water. Symptoms include bloody or watery droppings, loss of appetite, puffed-up feathers, and weakness. To prevent it, keep the brooder clean and dry, change bedding frequently, and provide good ventilation.

Natural preventatives like adding apple cider vinegar to water or feeding oregano and garlic can help strengthen immunity.[27] If symptoms appear, isolate the sick chick and use a coccidiostat treatment or herbal alternatives recommended by a vet or poultry expert.

Pasty Butt

A messy but common early problem, pasty butt happens when droppings dry and stick to a chick's vent, blocking the passage of waste. It's often caused by stress during shipping or other factors and temperature fluctuations. Check every chick daily for buildup and clean gently using a warm, damp cloth or cotton swab. Never pick at the dried excrement when it's dry. Applying a little olive oil or coconut oil around the vent can prevent recurrence.

Salmonella

While Salmonella is less common in home flocks, it's serious because it can spread to humans. Chicks can carry the bacteria sometimes without appearing sick. Common symptoms include diarrhea, droopiness, and dehydration.[27] Prevent the spread of infection by always washing your hands after handling birds, keeping their water and feed clean, and discouraging them from walking in their food bowls.

Keeping a close eye on your little ones' health, maintaining cleanliness, and providing natural immunity boosters will give them the best start in life, transforming the tiny, peeping fluffballs into plump, happy hens.

Brooder Cleaning

A clean brooder is the foundation of healthy chicks. Since young birds are especially vulnerable to bacteria and parasites, daily spot cleaning is essential. Remove damp bedding, spilled feed, and droppings every day, and fully replace all bedding weekly. Pine shavings or chopped straw work best because they're absorbent and soft under tiny feet. Between every new batch of chicks, scrub the brooder with warm, soapy water and disinfect it using a mild vinegar or poultry-safe cleaner to eliminate lingering germs.

Preparing Your Flock For the Future

As your chicks grow, you'll notice them becoming more curious and active. This is the perfect time to enrich their environment. Add small roosting bars or branches a few inches off the ground to help them practice perching. Toss in safe toys like hanging vegetables, shiny keys, or mirrors to prevent boredom and encourage natural behaviors such as pecking and exploring.

Once they are about 3–4 weeks old and fully feathered on their chests, they can enjoy short supervised trips outdoors in warm, dry weather. Use a secure, shaded pen to protect them from predators and drafts. Outdoor time strengthens their muscles, boosts immunity, and helps them adjust to new sights and sounds.

When it's time to move your chicks into the main coop, introduce them gradually. Allow them to see older hens through a fence first, then slowly integrate under supervision to prevent bullying. Size differences can trigger pecking order disputes, so patience is key. We'll explore the transition process and how to ensure a peaceful flock in greater detail in the next chapter so keep on reading.

From Egg to Chick: Hatching at Home

If you decide to begin your chicken-keeping journey from the eggs themselves, then you've got two exciting paths to choose from. You can let a broody hen do what nature designed her for, or take matters into your own hands with an incubator. Each method has its own kind of magic: the gentle clucking of a mother hen turning her eggs with care, or the thrill of watching life begin under the warm glow of an incubator. Whichever route you pick, the process is easier (and far more fascinating) than you might think once you learn the basics.

1. Choosing the Right Eggs

The journey to raising happy, productive hens begins with choosing the right eggs. Start by sourcing fertile eggs from a trusted breeder or a healthy flock with a strong, active rooster. Fertile eggs are the only ones capable of developing into chicks, so make sure the seller confirms fertility rather than simply offering table eggs. Look for signs of a well-managed flock: bright-eyed, clean birds with shiny feathers and no signs of illness or overcrowding. Healthy parent stock produces strong, resilient chicks.

When selecting individual eggs, pick medium-sized ones with smooth, uncracked shells. Oversized or undersized eggs can lead to weak or deformed chicks, while thin-shelled or misshapen eggs may not survive the incubation process. The shell's integrity is vital since it regulates air and moisture exchange during development.

Resist the urge to wash hatching eggs even if they have a bit of dirt on them. Washing removes the egg's natural protective bloom, a thin coating that helps keep bacteria out. Instead, gently brush away any debris with a dry, soft cloth if needed. Store the chosen eggs pointed-end down in a cool (not cold) place, around 12–15°C (55–60°F), until you're ready to start incubation.

2. The 21-Day Incubation Timeline

Whether it's under a hen or in a machine, hatching takes about three weeks. Keep these key milestones handy:

Day	What to Do	What's Happening Inside
Day 1	Set eggs in the incubator at 37.5 °C / 99.5 °F with 45 % humidity. Turn eggs 3–5 times daily.	Embryo begins forming.
Day 7	Candle the eggs (shine a small light through the shell). Remove any clear or infertile ones.	Veins and a tiny dark spot appear.
Day 14	Candle again. You should see movement! Continue turning.	Chick grows rapidly.
Day 18	Stop turning the eggs and raise humidity to 60–65 %.	Chicks position themselves for hatching.
Day 21	Hatch day! Leave the incubator closed until chicks are fully dry and fluffy.	Chicks pip, zip, and emerge.

Keep the incubator in a draft-free room and monitor temperature daily, even small fluctuations can delay hatching.

Quick troubleshooting:
If chicks don't hatch by Day 22–23, candle before discarding, late hatches can happen. Don't help chicks out of shells unless absolutely necessary, patience is part of the process!

Day 7: The First Peek Inside

At around seven days of incubation, it's time for your first candling session: a fascinating glimpse at what's happening inside the egg. In a dark room, use a bright candler or a strong flashlight to illuminate each egg.

- **Fertile Egg:** You should see a small, dark spot (the embryo) surrounded by a network of red, branching veins spreading outward like delicate roots. The air sac will be visible at the larger end of the egg, and you may even notice slight movement if the chick is active. This is a great sign that development is progressing normally.

- **Unfertilized or Non-Developing Egg:** The egg will appear mostly clear or with only a faint yolk shadow. There will be no visible blood vessels or embryo. If you see a faint red ring (called a blood ring), it means the embryo started to develop but stopped early.

Day 14: Watching Growth Take Shape

By two weeks, the developing chick fills most of the egg, making it appear much darker when candled.

- **Fertile Egg:** The egg will look mostly opaque except for a lighter oval at the large end, where the air sac sits. The chick's body now occupies most of the interior, and you might glimpse its eye, shape, or even movement if you look closely.

- **Unfertilized or Stopped Development:** The egg will still appear mostly clear or unchanged from day 7. A visible blood ring or cloudy area may indicate that the embryo died earlier in development.

- **Important Candling Tips:** Handle eggs carefully and keep sessions brief to avoid cooling them.

- Stop candling after day 18, as embryos are delicate and preparing for hatching.

- Always candle in a dark room using a bright, focused light.

- Remove any clearly unviable eggs to prevent bacterial contamination or explosions in the incubator.

- If you're unsure, it's perfectly fine to wait a day or two before rechecking, as long as it's before day 18.

3. Letting a Broody Hen Do the Work

If one of your hens decides to settle on a nest and refuses to move, congratulations, you've got a broody! This natural instinct is a gift, as she'll take over the hard work of incubating and turning the eggs for you. Place 6–10 fertile eggs under her in a quiet, secure corner of the coop where she feels safe from disturbance. Mark each egg lightly with a pencil so you can tell them apart and remove any new ones other hens may try to add.

Make sure your broody hen has food and water nearby, along with a clean, dry, and well-padded nest. Privacy is key. Too much disruption can cause her to abandon the clutch. A good broody will keep the eggs warm and humid, turning them several times a day without any help from you.

4. After the Hatch: Welcoming the New Chicks

Once the chicks begin to hatch, patience is important. Leave them under the hen or inside the incubator until they are completely dry, fluffy, and steady on their feet. This usually takes a few hours. Moving them too early can chill or weaken them.

Transfer the dry chicks to a warm, draft-free brooder lined with clean bedding and equipped with a reliable heat source. Maintain the temperature at 32–35°C (90–95°F) for the first week, then lower it by about 3°C (5°F) each week as they grow and develop feathers. Always give them enough space to move toward or away from the heat, so they can regulate their own comfort.

Temperature guide for first weeks:	
Week	Brooder Temperature
1	32–35°C (90–95°F)
2	29–32°C (85–90°F)
3	26–29°C (80–85°F)
4	Gradually reduce until fully feathered

5. Watching New Life Begin

There's nothing quite like the thrill of hatching your own chicks. It's a mix of wonder, excitement, and pure joy; a front-row seat to nature's little miracle. The days of waiting teach patience and care, and when you finally hear that faint peep from inside the shell, it's unforgettable. Watching a chick push its way into the world, damp and wobbly but full of life, reminds you just how amazing the cycle of life really is. It's one of those rare moments in chicken-keeping that never loses its magic, no matter how many times you experience it.

The Takeaway

Raising baby chicks is one of the most rewarding parts of keeping chickens. From the first peeps in the brooder to their first curious steps outdoors, every stage teaches you something new about nurturing life. With the right setup, clean environment, balanced diet, and a watchful eye for early signs of illness, your chicks will grow into healthy, confident birds ready to join the flock.

Beyond the basics of feed, warmth, and hygiene, raising chicks is also about building trust and routine. The time you spend handling them gently, observing their behavior, and ensuring their comfort lays the foundation for a calm, friendly flock later on. Each small act of care, whether sprinkling grit, adding a roost, or cleaning the brooder, creates a safer, happier home for your growing birds.

In the next chapter, we'll learn about the best ways to help your chicks transition to the coop. You'll learn a number of techniques to help your feather babies bond with your existing flock. This is where chicken society truly takes shape, the fascinating (and sometimes feisty) world of the pecking order. So let's move forward to the next chapter and see how to keep the peace in your growing flock!

Chapter 7

The Pecking Order

I still remember the day I decided my first batch of chicks was "ready" to join the big girls in the coop. I opened the gate, heart full of pride and within minutes, chaos broke out. A loud squawk, a flurry of feathers, and one of my gentle hens suddenly turned into a tiny feathered tyrant, chasing the newcomers around like they'd trespassed on royal ground. My heart sank. I stood there, horrified, feeling like I'd betrayed both my old hens and my babies.

That night, I scoured chicken forums and called a neighbor who'd raised birds for decades. Her laughter softened my panic. This was normal, she said. It was the pecking order, a natural part of flock life.

That moment changed everything for me. Understanding how chickens interact is key to keeping your flock healthy and stress-free. In this chapter, we'll explore the pecking order, roosters, broody hens, and peaceful introductions.

What Is the Pecking Order?

The term "pecking order" points toward the social hierarchy that determines who gets to eat first, roost highest, and lead the flock. This system keeps order among chickens and prevents constant chaos, maintaining an organized structure for food, space, and safety. Every flock, big or small, develops its own unique chain of command.

At the top stands the Alpha Hen, confident and assertive. She eats first, chooses the best nesting spot, and often leads the flock when foraging. The Beta Hens form the middle ranks. They respect the alpha but assert dominance over lower hens. At the bottom sits the Omega Hen, often the youngest or most timid, who eats last and avoids confrontation. Roosters, if present, usually sit above the hens in the hierarchy, managing disputes and keeping order.

The pecking order forms early. Chicks start sorting out dominance within weeks of hatching. In flocks raised together, this happens naturally with minor squabbles. But when new birds are introduced, things can get tense. Pecking, chasing, or brief fights are normal as everyone renegotiates their position.

When aggression turns relentless, like one hen bullying another or injuring other birds in the flock, it's time to step in. Such aggressive displays of dominance are usually a result of overcrowding, lack of food or space, or boredom. Separating the bully, adding distractions like hanging cabbages, or expanding the run are some ways to help restore peace.

One of the most common times pecking order issues flare up is when new birds are added to the flock. With a bit of planning, you can make those introductions far less stressful for everyone.

Integrating New Birds – A Step by Step Survival Guide

I'll never forget the day I tried to introduce three new pullets to my existing flock of five hens. I'd read somewhere that you should "just put them in and let them work it out," so that's exactly what I did. Within seconds, chaos erupted. My normally docile Buff Orpington turned into a feathered fury, chasing the newcomers around the run while the others joined in like a gang of schoolyard bullies.

The new girls huddled in a corner, terrified and trembling. One was bleeding from a peck to the comb. I felt like the worst chicken keeper in the world.

That disaster taught me a valuable lesson: introducing new birds isn't something you wing (pun intended). It requires patience, strategy, and a solid plan. Done right, integration can be smooth and stress-free for everyone involved. Done wrong, it can lead to injury, illness, and a fractured flock.

Let's walk through the process step by step so you can avoid my mistakes.

Why Integration Is So Important

Chickens are deeply social creatures with strict hierarchies. When a new bird enters the scene, the entire pecking order gets disrupted. Existing hens see newcomers as intruders and often react with aggression to defend their territory, resources, and status.

Without proper introduction protocols, you risk:

- **Injuries:** Pecking, chasing, and fighting can lead to wounds, infections, or even death.
- **Disease transmission:** New birds may carry illnesses your flock hasn't been exposed to.
- **Chronic stress:** Bullied birds may never fully integrate, leading to poor health and low egg production.
- **Flock division:** Instead of one cohesive group, you end up with cliques and constant tension.

The good news? With the right approach, you can minimize conflict and create a peaceful, unified flock.

The Golden Rule: Quarantine First

Before you even think about introductions, **quarantine new birds for 2-4 weeks.** This is non-negotiable.

Why Quarantine Matters:

- New birds may carry diseases (respiratory infections, parasites, viruses) that can devastate your entire flock.
- Stress from transport or rehoming can cause latent illnesses to flare up.
- Quarantine gives you time to observe the newcomers' behavior, appetite, and droppings.

How to Quarantine:

1. **Set up a separate space** at least 30 feet from your existing coop (a garage, shed, or spare coop works).
2. **Provide food, water, shelter, and enrichment** just like a regular coop.
3. **Wear dedicated clothing and shoes** when handling quarantined birds, and always tend to your main flock *first* to avoid cross-contamination.
4. **Watch for symptoms:** Sneezing, coughing, nasal discharge, lethargy, abnormal droppings, or sudden weight loss are red flags.
5. **Deworm and treat for parasites** during quarantine if needed.

If the new birds show *any* signs of illness, extend quarantine and consult a vet before proceeding with integration.

Step-by-Step Integration Process

Once quarantine is complete and your new birds are healthy, it's time to start introductions. This process can take anywhere from 1-4 weeks depending on flock dynamics.

Phase 1: See But Don't Touch (Days 1-7)

The goal here is to let the two groups get used to each other's presence *without physical contact*.

How to Do It:

1. **Place the new birds in a secure pen or cage *inside* or directly next to the main run.** Use wire fencing or hardware cloth so both groups can see, hear, and smell each other but can't make physical contact.

2. **Keep them separated for at least 3-7 days.** This allows the flock to acknowledge the newcomers without feeling threatened.

3. **Feed both groups near the divider** so they associate each other with positive experiences (food!).

What to Watch For:

- Curiosity (hens peering through the fence)
- Posturing (puffing up, head-bobbing)
- Minimal aggression (pecking at the fence is okay; relentless attacks are not)

Red Flag: If your existing flock is *constantly* attacking the fence or if the new birds seem excessively stressed (hiding, refusing to eat), extend this phase a few more days.

Phase 2: Supervised Free-Range Time (Days 8-14)

Once both groups seem calmer around each other, it's time for supervised introductions in a neutral space.

How to Do It:

1. Let both groups free-range together in a large, neutral area** (your yard, a pasture, or an expanded run). A bigger space reduces territorial behavior.

2. Choose late afternoon or early evening when chickens are naturally calmer and focused on foraging before roosting.

3. Stay present and watchful. Bring a broom, water bottle, or spray bottle to break up any serious fights (a quick spritz of water usually does the trick).

4. Provide multiple food and water stations so hens don't have to compete for resources.

5. Add distractions: Scatter scratch grains, hang cabbage or lettuce, or toss in some mealworms to keep everyone busy and redirect aggressive energy.

Duration: Start with 30-60 minutes and gradually increase over several days.

What to Watch For:

- **Normal**: Some chasing, pecking, posturing, and squabbling. The new birds will likely stay together and avoid the established flock at first.
- **Concerning**: Relentless chasing, cornering, ganging up on one bird, or blood drawn. If this happens, separate them immediately and try again the next day.

Pro Tip: Introducing new birds at dusk can help! Place the newcomers on the roost after dark when everyone's sleepy. Chickens have terrible night vision, so by morning, they may accept the new birds as part of the flock. This works best with calm, confident flocks.

Phase 3: Nighttime Cohabitation (Days 10-14)

If supervised free-range sessions go well, it's time to let them share the coop overnight.

How to Do It:

1. After a successful supervised session, herd everyone into the coop together at dusk.
2. Place the new birds on the roost (not in a nesting box, which can trigger territorial behavior).
3. Close the coop door and let them sleep together.
4. Check on them first thing in the morning. Open the coop early so everyone can escape into the run if tensions rise.

What to Watch For:

- The first few mornings may involve some squabbling as the flock reestablishes order.
- Watch for birds being blocked from food/water or cornered. If one hen is being relentlessly bullied, separate her and try again in a few days.

Phase 4: Full Integration (Week 2-4)

By now, the flock should be settling into a new pecking order. Continue monitoring daily, but give them space to work things out.

What Success Looks Like:

- All birds roosting together at night
- Minimal chasing or pecking (brief squabbles are normal)
- New birds eating, drinking, and foraging confidently
- No visible injuries or signs of extreme stress

What to Do If Problems Persist:

- **Provide more space:** Overcrowding intensifies aggression.
- **Add visual barriers:** Branches, tarps, or partitions create hiding spots and reduce line-of-sight aggression.
- **Remove the bully:** If one hen is the main aggressor, isolate her for 3-5 days. When reintroduced, she'll have lost some status and may be less aggressive.
- **Rehome if necessary:** In rare cases, some birds simply won't integrate. If a hen is severely injured or chronically stressed, rehoming may be the kindest option.

Special Considerations

1. **Age Differences:**
   - **Chicks**: NEVER introduce young chicks (under 12 weeks) directly to adult hens. Wait until they're fully grown and similar in size, or the adults may seriously injure or kill them.
   - **Pullets**: Young pullets (4-6 months) can be integrated but may be picked on more. Follow the full integration process carefully.
   - **Mature hens:** Adult hens of similar age integrate most smoothly.

2. **Size Differences:**
   - **Bantams + Standard hens:** Mixing sizes can work, but watch closely. Larger birds may bully smaller ones, and roosters may injure bantam hens during mating.

3. **Rooster Introductions:**
   - Introducing a new rooster is *extremely* tricky and often leads to violent fights. If you must do it, remove the existing rooster first, let the new one settle in, then reintroduce the old rooster using the same phased approach. Be prepared for ongoing conflict and have a backup plan (separate living quarters or rehoming).

4. **Broody Hens:**
   - Broody hens with chicks can be integrated after the chicks are 6-8 weeks old. The mother will defend her babies, but adult hens may still attack chicks. Use a separate pen-within-a-pen setup initially.

Tips for Smooth Integration

→ Introduce multiple birds at once. A lone newcomer will be bullied more than a pair or trio who can support each other.

- → Match sizes and ages. The closer the birds are in size and maturity, the easier integration will be.
- → Introduce during spring or summer. Longer daylight hours and access to outdoor space reduce stress. Avoid winter introductions when hens are cooped up and irritable.
- → Rearrange the coop. Move roosts, feeders, and nesting boxes during integration. This "resets" the territory and prevents established hens from guarding favorite spots.
- → Add extra resources. More feeders, waterers, roosts, and nesting boxes reduce competition.
- → Use "integration accessories." Some keepers swear by:
  - Chicken saddles (protective vests) on newcomers to prevent feather damage
  - Pinless peepers (plastic blinders) on aggressive hens to reduce pecking accuracy
  - Rooster collars to quiet overly vocal roosters during integration stress
- → Be patient. Integration takes time. Rushing the process leads to injury and setbacks.

When to Step In vs. Let Them Sort It Out

It's hard to watch your birds squabble, but some conflict is normal and necessary. Here's how to tell the difference between "working it out" and "dangerous aggression":

Let Them Sort It Out:
- Chest bumping, wing flaring, head-bobbing
- Brief chases followed by the underdog retreating
- Pecking that doesn't draw blood
- Posturing and vocalizations

Step In Immediately:

- Prolonged, relentless chasing (more than 30 seconds)
- Ganging up (multiple hens attacking one)
- Pecking that draws blood, especially around the head/comb
- A bird being cornered with no escape route
- Signs of extreme distress (panting, trembling, refusing to move)

My Integration Success Story

After my disastrous first attempt, I learned my lesson. The next time I added new birds, I followed the phased approach religiously. I quarantined three ISA Brown pullets for three weeks, then set up a wire pen inside the run for "see but don't touch" time.

After a week, I let everyone free-range together under supervision. There was some posturing and chasing, but nothing serious. I scattered mealworms to distract them, and within 20 minutes, everyone was foraging peacefully (mostly because they were too busy eating to fight).

That night, I placed the new girls on the roost after dark. By morning, the flock had accepted them as part of the crew. There were a few minor squabbles over the next week, but nothing alarming. Within two weeks, the pecking order had stabilized, and my flock was harmonious again.

It took patience and planning, but it worked. And let me tell you, watching all eight hens dust-bathe together in the afternoon sun made every bit of effort worth it.

Integration Checklist

- ☐ Quarantine new birds for 2-4 weeks
- ☐ Observe for illness, parasites, or abnormal behavior
- ☐ Set up "see but don't touch" pen inside/next to main run
- ☐ Keep groups separated for 3-7 days
- ☐ Start supervised free-range sessions in a neutral area
- ☐ Provide multiple food/water stations and distractions
- ☐ Gradually increase free-range time over several days
- ☐ Place new birds on roost at dusk for overnight cohabitation
- ☐ Monitor closely for the first 1-2 weeks
- ☐ Rearrange coop to "reset" territory
- ☐ Be patient and intervene only when necessary
- ☐ Celebrate when everyone's happily coexisting!

The Bottom Line

Introducing new birds doesn't have to be a bloodbath. With quarantine, phased introductions, and plenty of patience, you can expand your flock peacefully and safely. Yes, there will be some squabbling, that's chicken society at work. But if you follow these steps, you'll minimize stress, prevent injuries, and end up with a happy, unified flock. And when you see your old hens and new girls roosting side by side, you'll know it was all worth it.

With introductions behind you, it's helpful to know what normal, healthy flock behavior looks like. Recognizing these cues will make you a more confident, responsive chicken keeper.

Happy Flock, Happy Life: Reading Chicken Behavior

Chickens may not talk, but they're fantastic communicators once you know what to look for. Understanding body language and social cues can save you stress (and a few feathers).

Common Chicken Behaviors	
Behavior	Meaning / What It Tells You
Raised hackles or chest bumping	A power struggle or dominance display. Hens or roosters are establishing the pecking order. Brief squabbles are normal unless they turn violent.
Head bobbing or crouching	A sign of submission or respect toward a dominant bird. Indicates that social order is stable and the flock is calm.
Feather pecking or bullying	A warning sign of stress, boredom, or overcrowding. Provide more space, distractions, and foraging opportunities to reduce tension.
Gentle pecking	Part of normal flock interaction. Chickens use light pecks to explore, show curiosity, or maintain mild social bonds.
Loud clucking or squawking	Can signal alarm, excitement, or a predator nearby. Investigate if it sounds persistent or panicked.
Soft murmuring or contented clucks	Indicates relaxation and comfort — your hens are happy and feel safe in their environment.
Pacing or restlessness	Suggests discomfort, heat stress, or lack of stimulation. Check for ventilation, shade, or boredom.

You can gently train hens using repetition and positive reinforcement by talking softly, feeding them by hand, and being patient. Calm hens are friendlier, easier to inspect, and happier to follow your lead.

Brooders and Roosters

Brooders and roosters play very different but equally important parts in the life of a flock. The broody hen is nature's incubator, sitting patiently on her eggs until they hatch. The rooster, on the other hand, stands guard, keeps order, and ensures the continuation of the flock. Knowing how they think and behave helps you manage your chickens with less chaos and more success.

Brooding Hens

A broody hen is one that's overcome with its motherly instinct to hatch eggs. She'll fluff up her feathers, sit on the nest for long hours, and may even hiss or peck when you come near. A broody hen stops laying eggs, eats less, and can overheat in warm weather. If you're not planning to hatch chicks, it's best to gently break her broodiness by cooling her underside, removing tempting nesting spots, or placing her in a wire-bottom cage with good airflow.

But if you do want chicks, a broody hen can be nature's best incubator. Move her carefully to a separate brooding box in a quiet, safe corner. This is known as the "Box-In" method, which helps the hens stay committed to their nest, keeping the eggs safe. Expect her to be defensive for the first 24 hours, then settle in. Provide food and water close by, as she'll rarely leave the nest.

Not all breeds go broody though. Leghorns, Polish, and Hamburgs almost never do. Silkies and Orpingtons, on the other hand, are famously maternal. Whether you're discouraging or encouraging broodiness, understanding your hen's instincts and responding to its needs is the key to keeping a happy flock.

Roosters

Roosters often get a bad reputation. They're seen as noisy, aggressive, and unnecessary troublemakers. But once you start observing them closely, you realize how misunderstood they really are. Roosters are intelligent, sensitive, and deeply social animals who play a vital role in the dynamics of the flock. They're not just the loud ones that wake you up at dawn; they're protectors, providers, and even peacekeepers when given the right environment. Let's look at some myths about roosters and learn what they're really like.

Myth Number 1: They're Not That Smart

FACT: Chickens are highly sensitive and intelligent beings. Roosters are no different.

Many people underestimate chickens, thinking they just peck around without much thought. But research and real-life observation show just how clever these birds can be. Their excellent memory helps them recognize up to a hundred individual faces, both human and avian. They can distinguish colors and shapes, use logic to solve problems, and even demonstrate emotional intelligence by comforting one another in stressful situations and learning from each other's experiences.

Their senses are far more refined than we imagine. They can see ultraviolet light, detect tiny movements, and hear frequencies beyond human range. They notice tone and body language, which is why your energy affects how they respond to you.

Once I spent a week watching my rooster, Rusty, communicate with his hens. He'd cluck softly when finding food, call out alarms when a hawk passed overhead, and lower his voice when the flock rested. His behavior wasn't random; it was strategic, protective, and deeply attuned to his surroundings.

Myth Number 2: They Have No Personality

FACT: Every chicken has a distinct, unique personality, including roosters!

Just like humans, no two chickens are alike. Some roosters are bold and daring, while others are gentle and cautious. One might strut confidently around the coop like a showman, while another quietly follows his hens, letting them eat first. Some are curious, testing every new object, while others hang back, observant but reserved.

Chickens even have emotional temperaments. Some are playful, others broody or contemplative. Roosters, especially, show incredible social awareness. They mediate disputes between hens, break up squabbles, and sometimes even stand guard over weaker members of the flock. In short, they have as much personality as your pet dog. They just express themselves differently, often flapping their wings, fluffing up their feathers, and clucking loudly to show their joy or displeasure.

Myth Number 3: They Are Loud, Inherently Aggressive, and Dangerous

FACT: A rooster is the hero his family deserves. A watchful protector who won't let anyone harm his flock.

Roosters crow for many reasons. Crowing is their form of communication. It's a territorial signal, a greeting, or even a response to environmental cues like changes in light or sound. In nature, a rooster's crow tells predators: "I'm watching. This flock is protected."

In a healthy flock, a rooster maintains order, not chaos. He ensures his hens have access to food, warns them of danger, and sometimes lets them know when it's safe to relax. His aggression toward other roosters often stems from instinct. He must defend his role as the family's guardian. If space is limited or the environment stressful, this behavior intensifies. But in large enough spaces with balanced flock ratios, roosters coexist surprisingly well.

They sometimes get rough with hens during mating, but it's important to recognize the natural dynamics at play. Overbreeding or stress can intensify this behavior, but with proper management such as providing enough hens per rooster, and no overcrowding these issues get resolved.

Most aggressive roosters are simply fearful or territorial. They're reacting to perceived threats. Building trust early, respecting their space, and avoiding sudden movements can prevent defensive behavior. People tend to have a lot of misconceptions about these birds. Now that we've dispelled some of these, let's address some common questions about roosters and understand what their different behaviors mean.

Why Roosters Rumble?

Believe it or not, when two roosters fight, it's not mindless violence. It's their way of communicating with each other. They fight each other to establish hierarchy. They'll puff up, circle each other, and may clash briefly before one backs down. This display sets boundaries and reinforces social balance. Problems arise when there's overcrowding or too few hens. These conditions push roosters beyond natural competition into chronic conflict. Providing adequate space, visual barriers, and multiple feeding stations helps prevent such disputes.

Three main reasons roosters brawl are:

1. Dominance: They're establishing who's in charge.

2. Mating rights: Competition over hens, especially in small spaces.

3. Resource guarding: Protecting food, shelter, or preferred perches.

When it comes to balance, one rooster for every 8–12 hens is ideal. Too few hens can lead to overmating and stress, while too many may leave him overwhelmed and unable to protect or manage his flock effectively.

In flocks with multiple roosters, they often sort out their hierarchy naturally once they mature. This process usually involves sparring, which are brief, non-lethal scuffles where they flap, peck, or jump to assert dominance. These encounters may look dramatic but are rarely dangerous. Once the pecking order is established, peace typically returns. Providing plenty of space ensures these natural dominance displays don't escalate into real fights.

Pros and Cons of Keeping Roosters

A rooster is the flock's natural guardian, always on alert for predators. If you want fertile eggs and natural reproduction, a rooster is essential. His presence brings order and security to the flock, keeping everyone in line and easing the introduction of new hens. And while not everyone loves an early morning crow, some chicken keepers appreciate the old-time alarm clock. Lastly, roosters add diversity and personality to the coop. Each one unique in temperament and beauty, often becoming the farm's most charismatic resident.

Of course, keeping these territorial birds comes with challenges. Some can be rough on hens, overmating or pulling feathers during breeding. Fighting may occur when multiple roosters compete for dominance, especially in cramped environments. Their crowing, though charming to some, can contribute to noise pollution in suburban or urban areas. All this makes it essential for you to check your local laws before bringing one home because many towns restrict or ban roosters due to noise complaints.

While roosters and hens bring so much life and personality to your coop, there's another side to chicken keeping that every caretaker eventually faces. It's not a fun topic, but it's an important one: knowing what to do when a bird's life nears its end.

Chapter 8.

The Hardest Part: End-of-Life Decisions

This is the chapter I didn't want to write. But it's also the one I wish someone had written for me when I needed it most.

Her name was Dorothy, the first hen I ever owned, my loud, sassy girl who followed me everywhere and bossed the entire flock around like a tiny feathered dictator. She lived a good, long life: five years of strutting, dust-bathing, and laying beautiful brown eggs. But that last year, she slowed down. Her comb paled. She stopped laying. She spent more time sitting than scratching.

Then one morning, I found her struggling to breathe. Her eyes were dull, and she barely moved when I picked her up. I knew. Deep down, I knew.

But knowing and accepting are two different things. Chickens don't live forever. As much as we wish they could, there comes a time when age, illness, or injury means they're suffering, and the kindest thing we can do is let them go peacefully.

This section isn't easy to read, and it wasn't easy to write. But if you love your birds, you owe them this final act of compassion.

Recognizing When It's Time

One of the hardest parts of chicken keeping is knowing when to intervene. Chickens are stoic, they hide pain and illness as a survival instinct. By the time they show obvious symptoms, they're often far worse off than they appear.

Here are the signs that a hen may be suffering and unlikely to recover:

Physical Signs:

- **Severe weight loss** despite eating (you can feel her keel bone protruding sharply)
- **Inability to stand, walk, or move** (paralysis, severe leg injuries, neurological damage)
- **Labored breathing** that doesn't improve with rest or treatment
- **Prolapsed vent** that can't be corrected
- **Chronic, untreatable illness** (advanced tumors, severe organ failure, infectious diseases that don't respond to antibiotics)
- **Severe injuries** from predator attacks that cause unmanageable pain or permanent disability

Behavioral Signs:

- **Isolating from the flock** and showing no interest in food, water, or social interaction
- **Unresponsive or lethargic** even when approached
- **Visible distress:** panting, gasping, trembling, repeated vocalizations of pain
- **Loss of balance or coordination** with no improvement

Quality of Life Questions:

Ask yourself:

1. Is she eating, drinking, and moving on her own?

2. Is she engaging with the flock or showing interest in her surroundings?

3. Is she in constant pain despite treatment?

4. Has her condition improved, stabilized, or worsened over the past few days/weeks?

5. Can this condition realistically be treated or managed, or is it terminal?

If the answer to most of these questions points toward suffering with no hope of recovery, it may be time to make the difficult decision.

Consulting a Vet

Before making any final decisions, consult an avian or poultry veterinarian if possible. They can:

- Accurately diagnose the problem
- Provide treatment options you may not have considered
- Offer an honest assessment of the hen's quality of life and prognosis
- Perform humane euthanasia if needed

Not every condition is a death sentence. Sometimes, what looks dire can be treated successfully. But if a vet confirms that euthanasia is the most humane option, trust their expertise.

Humane Euthanasia: Your Options

When the time comes to help a suffering hen pass peacefully, the goal is the same: to end pain quickly, humanely, and with dignity. You have several options, and the right choice depends on your location, resources, and comfort level.

Veterinary Euthanasia (Recommended)

This is the most peaceful and stress-free option for both you and your hen. A veterinarian will:

1. Administer a sedative to help your hen relax
2. Follow with a painless injection that stops the heart within seconds
3. Allow you to be present, hold her, and say goodbye

Why This Is the Best Choice:

- Quick and completely painless
- No risk of error or prolonged suffering
- Professional handling reduces emotional burden
- You can focus on being present for your hen rather than performing the procedure yourself

Practical Considerations:

- Cost: Typically $50-$150+ depending on location
- Availability: Not all vets treat chickens—call ahead to confirm
- You can request a house call if your vet offers farm visits
- Some vets offer group rates if you need to euthanize multiple birds

Many keepers find that the peace of mind and gentle goodbye are worth the cost. If finances are tight, explain your situation—some vets offer sliding scale fees or payment plans.

At-Home Euthanasia Methods

For those in rural areas without access to avian veterinary care, or for experienced keepers comfortable with traditional farm practices, humane at-home methods do exist. These include:

- **Cervical dislocation** (quick dislocation of the neck)
- **Carbon dioxide (CO_2) chamber**
- **Decapitation** (using proper tools)

Each of these methods, when performed correctly, results in instantaneous unconsciousness and death without prolonged suffering. However, they require:

- Proper technique and training
- Emotional readiness
- Physical capability
- Confidence in execution

If you are considering an at-home method, please see Appendix B: Humane Euthanasia Methods for detailed, step-by-step instructions. That section includes safety guidelines, required equipment, and proper techniques to ensure the process is as swift and painless as possible.

Important Notes:

- **These methods are not for everyone**, and there is no shame in choosing veterinary care instead
- Improper technique can cause suffering, if you're uncertain, seek professional help
- **Never use** drowning, suffocation (without proper CO_2), freezing, or blunt force trauma—these cause prolonged suffering and are inhumane
- Consult with experienced mentors, agricultural extension offices, or homesteading communities for guidance if needed

The decision to perform at-home euthanasia is deeply personal. Some keepers find it a final act of care and closure. Others prefer the expertise and emotional distance of veterinary services. Neither choice is wrong—what matters is that your hen's suffering ends quickly and peacefully.

After Euthanasia: Saying Goodbye

Once your hen has passed, take time to process your emotions. This is a real loss, and it's okay to grieve.

Coping with Grief

Losing a hen, especially one you've bonded with, is painful. Don't let anyone tell you "it's just a chicken." Your grief is valid.

Ways to Honor Your Hen:

- Plant a tree, flower, or herb in her memory
- Write about your favorite memories in a journal
- Create a small memorial stone or marker in your garden
- Donate to a poultry rescue or animal sanctuary in her name
- Frame a favorite photo
- Allow yourself time to heal before bringing in new birds

Talk to Others: Chicken-keeping communities (online and local) understand this loss. Reach out to fellow keepers who've been through it, you're not alone. Many find comfort in sharing stories and supporting each other through these difficult decisions. If you need gentle support through this grief, there are audio sessions at www.easevibes.com designed specifically for grief and pet loss that can offer comfort during the hardest days.

Give Yourself Permission:

- To cry
- To feel silly for grieving "just a chicken" (you're not silly—your feelings are real)
- To take a break from the coop for a day or two
- To remember the joy she brought you, not just the sadness of loss

When the Flock Mourns

Chickens notice when a flock member is gone. You may observe:

- Increased vocalizations (calling for the missing hen)
- Restlessness or searching behavior
- Changes in pecking order as remaining hens adjust
- Some hens may seem subdued or quieter than usual

How to Help Your Flock:

- Maintain regular routines (feeding, bedtime, free-range time)
- Provide extra treats and enrichment
- Spend time with them—your presence is comforting
- Give them time to adjust (usually a few days to a week)
- Avoid introducing new birds immediately (let the flock stabilize first)

Dorothy's Last Day

When I finally made the decision about Dorothy, I called my vet. She came to the house that afternoon, and I held Dorothy in my arms while the vet administered the sedative. Dorothy's eyes closed softly, her breathing slowed, and then she was gone. It was over in seconds, no pain, no fear, just a quiet, peaceful release.

I buried her under the apple tree in my backyard, the same tree where she loved to forage for bugs on sunny afternoons. I wrapped her in an old dish towel—one she'd "helped" me fold countless times by standing on it, and placed her in a wooden box my neighbor had built for me.

Every spring when that tree blooms, I think of her. I remember her bossy clucks, the way she'd peck at my shoelaces, and how she'd rush over whenever I called her name. She taught me so much: how to care for chickens, how to laugh at their antics, and ultimately, how to let go with love and grace.

You're Doing the Right Thing

If you're facing this decision, know that you're doing the right thing. It takes courage to end suffering. It takes love to put their needs above your own heartache.

You gave your hen a good life, fresh air, green grass, treats from your hand, and your voice calling her name. You protected her, cared for her, and when the time came, you made sure her suffering ended peacefully.

You're a good chicken keeper. And your hen is lucky to have you.

Raising Chickens for Meat: A Humane Approach

There's another aspect of end-of-life care that deserves honest discussion. Many chicken keepers choose to raise their birds not only for eggs but also for meat, and this practice is entirely valid when approached with responsibility and respect. The most important priority is to ensure that the animal's life ends humanely; carried out with empathy, calmness, and as little pain or stress as possible. Ethical handling, proper preparation, and a method that prioritizes the bird's welfare are essential. For those choosing to process their own chickens, a clear, humane procedure should always be followed; see Appendix B for a step-by-step method designed to minimize suffering and maintain dignity throughout the process.

Even with all the practical steps and ethical guidelines, nothing truly prepares you for the moment it becomes personal. It's one thing to understand the process in theory; it's another to face it with a bird you've raised and cared for.

End-of-Life Decision Checklist

☐ Observe for signs of suffering vs. temporary illness
☐ Consult an avian or poultry vet for diagnosis and guidance
☐ Ask quality-of-life questions honestly
☐ Choose a humane euthanasia method (veterinary or at-home)
☐ If choosing at-home method, review Appendix B for proper technique
☐ Arrange for aftercare (burial, cremation, or disposal)
☐ Allow yourself time to grieve

- ☐ Honor your hen's memory in a meaningful way
- ☐ Support remaining flock members as they adjust
- ☐ Remember: You gave her a good life, and a peaceful end

This chapter may feel heavy, but you've walked through it with the same courage and tenderness you bring to your flock every day. Doing what's right, especially when it hurts, shows the depth of your commitment as a keeper. In honoring a hen's final moments, you honor her entire life. And that is something to be proud of.

The Takeaway

From the earliest days in the brooder to the first introduction into the coop, raising chickens teaches you patience, observation, and respect for nature. Learning about the pecking order, broody hens, and rooster dynamics allows us to make sense of the chaos and understand the intricate social system within the flock built on communication, hierarchy, and care. Each squabble, crow, and egg is part of a larger story of connection between you and your chickens. And as that relationship deepens, you'll discover that your birds offer far more than just breakfast.

Once your flock settles in and you've got a handle on their behavior, you'll find they're not just limited to producing eggs. They're powerful allies in building a more sustainable, low-waste backyard system. In the next chapter, we'll explore the concept of raising chickens the sustainable way.

Chapter 9.

Co-Op With Your Coop

I remember the summer my garden nearly gave up on me. Slugs were devouring my lettuce, beetles chewed through the squash, and every eco-friendly remedy I tried seemed to make no difference at all. I was frustrated and felt helpless. Then, one afternoon, I forgot to latch the gate, and one of my hens strutted right in. Within minutes, she was darting through the beds, expertly snatching bugs I couldn't even see. By the next morning, the damage had slowed, and my curiosity sparked.

Watching them work with their focused eyes and determined pecks was oddly satisfying, almost magical. My garden began to heal, and I learned something important: nature works best when we let it.

In this chapter, we'll explore how to use your flock to naturally manage pests, enrich soil, and build a backyard system that thrives in harmony with the earth.

Catch'em And CAP'M

CAP'M, an abbreviation of Chicken-Assisted Pest Management (read "cap 'em!"), is nature's best pest control system. Your chickens are incredible gardeners, tackling bugs, weeds, and waste with their mighty appetites. When allowed to roam through garden beds, they feast on slugs, beetles, caterpillars, and even centipedes, keeping infestations under control without a drop of pesticide. But pest control is just part of their magic. As they scratch and peck, they gently aerate the soil, breaking up clumps and improving texture. Their droppings add a rich, organic fertilizer that boosts plant growth naturally.

Letting your chickens into the garden occasionally is a low-cost and efficient way to get rid of unwanted pests. Think of CAP'M as a partnership between feathered farmers and human gardeners. It saves time, effort, reduces waste, and keeps the ecosystem thriving, one peck at a time.

Step-by-Step Guide to CAP'M

Creating a garden that works with your chickens instead of against them takes a little planning, but the rewards are well worth it. A chicken-friendly garden not only keeps your flock happy and healthy but also puts their natural instincts to good use, helping your plants flourish. Let's look at a step-by-step guide to create a garden setup with your hens in charge.

Step 1: Protect Fragile Plants

Start by fencing off delicate plants in your garden with wire mesh or raised planters. Chickens love to dig and dust-bathe, which can uproot tender seedlings in seconds. Use garden cloches, row covers, or small decorative fences to protect young plants while still allowing the birds to roam nearby.

Step 2: Rotational Grazing in the Garden

Designate sections of your garden for chickens to forage in rotation. This gives plants time to recover while letting your birds snack on pests and weeds. Portable fencing or chicken tractors make this easy and prevent overgrazing.

Step 3: Keep Chickens Away from Specific Plants

Some fragile plants like tomatoes, peppers, and strawberries may not respond well to CAP'M. Block access to these plants or set up barriers around these areas. Also, make sure your chickens don't wander anywhere near toxic plants such as foxglove, nightshade, and rhubarb leaves.

Step 4: Provide Alternatives for Scratching and Pecking

While moderate scratching and pecking helps aerate the garden soil, excessive amounts can damage your plants. A great way to minimize damage and maximize benefits is to limit the number of chickens you let into the garden. A large flock is bound to cause more harm than good. You can also create an area near your garden patch where your girls can enjoy a nice and refreshing dust bath.

Simply fill the area with sand, wood ash, and diatomaceous earth, and scatter scratch grains. This channels their energy away from your prized garden beds. Another tip is to supervise your flock's free-ranging time, and introduce your flock to the garden gradually.

Beyond the Garden Beds: Expanding Your Flock's Natural Abilities

Your chickens can work for you in more places than the vegetable garden. Even if your flock is confined to a coop or run, you can still put their natural instincts to use. Let them scratch through compost piles or dig through garden debris, where insects and larvae often hide. Their constant turning and aeration help the compost break down faster while reducing pest populations at the source.

If your flock free-ranges, they can cover a much wider area, eating insects while trimming weeds and scattering manure for added soil nutrition. Moving chickens around the yard or garden prevents bare patches and gives each area time to recover. Targeted supervised time in pest-heavy zones, like areas overrun with slugs or beetles, lets them focus their efforts where you need help most.

Chickens are also valuable partners beyond the garden. In pastures, they help control flies and ticks around livestock by breaking up manure piles and hunting pests hiding in the grass. In orchards, they clean up fallen fruit, larvae, and insects before they can spread or begin the next generation.

Used strategically, your flock becomes a natural, chemical-free management system that keeps pests in check simply by allowing your birds to do what they love most: scratch, forage, and hunt for the next tasty bug.

Bug Banquet: Which Insects Do Chickens Eat

Your girls are enthusiastic insect hunters! They'll happily eat beetles, grubs, ants, termites, flies, earwigs, spiders, grasshoppers, slugs, snails, and even small centipedes. These protein-rich snacks keep your flock healthy while dramatically reducing garden pests. You may never see a beetle or slug agin in your yard after letting your chickens take the wheel.

While most insects are safe, avoid letting chickens eat toxic bugs such as blister beetles, fireflies, or brightly colored caterpillars like monarchs, which can be poisonous. Always research local pests before encouraging your flock to forage freely.

For extra pest control, encourage your chickens to forage under fruit trees or established plants where insects tend to gather. Active breeds like Leghorns, Rhode Island Reds, and Australorps excel at hunting, while Bantams are ideal for smaller gardens since they scratch less agressively.

Chicken-Assisted Composting

If there's one thing chickens excel at, besides eating bugs, it's turning scraps into gold. Chickens are natural composters, shredding food waste, spreading organic matter, and supercharging decomposition with their nitrogen-rich manure. Instead of tossing kitchen scraps straight into a compost pile, let your flock do the work first. This system, often called chicken-assisted composting, saves labor, reduces waste, and gives your garden a steady supply of nutrient-rich soil.

Collect veggie peels, fruit scraps, stale bread, and garden trimmings. Avoid anything moldy, salty, or oily. Toss the scraps into a compost run or designated bin where your hens can peck, scratch, and mix everything naturally. They'll eat what they like and churn the rest into the soil.

You'll need a compost area or run, some carbon material (like straw, dried leaves, or wood shavings), and a shovel for occasional turning. If space allows, create a compost rotation system when one pile is ready to rest, open the next for your flock to work on.

Chicken manure is one of the most potent organic fertilizers available, rich in nitrogen, phosphorus, and potassium. However, it's too strong to use fresh, as it can burn plants. Always let it compost for at least 3–6 months to mellow out.

Rake manure, bedding, and food scraps into small mounds inside the run. Your chickens' constant scratching keeps the pile aerated and evenly mixed, no pitchfork required. Once the compost is dark and crumbly, move it to your garden beds for use.

Step-by-Step Guide to Composting Chicken Manure

Turning chicken manure into usable compost is straightforward once you understand the basic process. The following steps break down how to handle, balance, and age your manure safely so it becomes a reliable, nutrient-rich addition to your garden.

Step 1: Understanding How to Compost Chicken Manure

Chicken manure is high in nitrogen, which can burn plants if applied fresh. Composting stabilizes these nutrients, making them safe and accessible. There are two main methods of composting.

- Hot composting speeds up decomposition by maintaining a temperature between 130 – 150°F (55 – 65°C). It kills pathogens and weed seeds but requires regular turning and moisture monitoring.
- Cold composting takes longer—6 to 12 months—but requires less effort.

Choose a shaded, well-drained location away from water sources. Gather materials like chicken manure (greens) and bedding such as straw, sawdust, or dried leaves (browns). Turn the pile every few days (hot) or every few weeks (cold) to keep it aerated. Maintain moisture similar to a wrung-out sponge, and check temperature with a compost thermometer. Once the pile cools and the contents turn dark, crumbly, and earthy-smelling, it's ready!

Step 2: Knowing What to Add to Chicken Manure for Composting

Getting the green/brown ratio right is key; aim for roughly 1 part manure (green) to 2–3 parts bedding (brown). A healthy compost pile should smell earthy. If it stinks like ammonia, you have too much nitrogen, add more carbon (browns). For cold composting, increase browns to slow the breakdown and reduce odor. For hot composting, balance the mix for fast heating and efficient decomposition.

Step 3: How Long Does Chicken Manure Take to Compost?

Typically, chicken manure takes 3 to 6 months to mature in a hot pile, and up to a year in a cold system. Always age manure thoroughly before use. The risk of using it too soon includes root burn, pathogen exposure (like Salmonella), and nitrogen overload, which can stunt growth instead of helping it.

Step 4: How to Use Chicken Manure Compost

Chicken manure compost enriches the soil with slow-release nutrients, encourages earthworms, and helps plants grow greener and stronger. Mix a 1–2 inch layer into the topsoil in the garden before planting or spread it as mulch around vegetables, flowers, and shrubs. In pots, blend one part compost with three parts potting soil to boost nutrient content without overwhelming roots.

Step 5: Make Chicken Manure Tea

To give your plants a quick nutrient boost, you can make chicken manure tea. Simply fill one-third of a bucket with composted manure. Add water, stir, and let it steep for 1–2 days. Strain the liquid and dilute it 1:4 with water. Pour this liquid around the base of plants (never on the leaves) to provide a gentle yet powerful fertilizer that supports fast, healthy growth.

Your chickens can transform waste into wealth, fueling a thriving, self-sustaining garden. Rich in nitrogen, phosphorus, and potassium, it turns kitchen scraps and coop cleanings into a valuable resource. When composted properly, it becomes part of a sustainable cycle that feeds your garden, which in turn can help feed your flock.

Level-Up: Bring in the Chicken Tractors

As we discussed in Chapter 3, a chicken tractor is a mobile chicken coop without a floor that can be moved around your yard or pasture. It gives chickens access to fresh grass, weeds, and bugs while keeping them safely enclosed from predators. The concept is simple but incredibly effective: your birds do the work of tilling, fertilizing, and pest-controlling while getting a constant supply of fresh forage.

So, why raise chickens on a pasture using a chicken tractor? For one, it leads to a healthier diet for your birds and for you. Chickens that graze on greens and insects produce richer eggs and more flavorful meat packed with omega-3 fatty acids and loads of vitamins. Secondly, chicken tractors boost soil health by spreading manure naturally and preventing nutrient buildup in one area. As the tractor moves, your chickens aerate the ground and leave behind a perfect layer of fertilizer for your garden or pasture.

Chicken tractors also protect your flock from predators. Since the birds are enclosed, they're safe from hawks, dogs, and foxes while still enjoying outdoor freedom. Finally, they reduce pest populations. As your flock moves through an area, they'll scratch up grubs, beetles, and larvae that might otherwise harm your plants.

Whether you have a small backyard garden or a farm, using a chicken tractor is an efficient way to make the most of your flock. With just a bit of daily movement, you'll create healthier birds, richer soil, and a cleaner coop environment.

How to Manage a Chicken Tractor

Managing a chicken tractor effectively means balancing mobility, safety, and comfort for your flock. Whether you're raising a few backyard hens or managing a small homestead, a well-designed and maintained chicken tractor helps your birds stay healthy, creates fertile soil, and minimizes your daily chores.

Let's start with designing your own mobile coop. Chicken tractors come in many styles: pasture pens (low, rectangular enclosures that sit close to the ground) and pasture shelters (taller, walk-in structures). Your choice depends on flock size, terrain, and how often you plan to move the tractor. Larger tractors allow more space but can be heavier and harder to move, while smaller ones are more portable but limit flock size.

When designing, consider the trade-offs between size, weight, and stability. Lightweight materials make moving easier but can reduce sturdiness during wind or storms. You can also add wheels or skids to make relocation simpler. This will allow you to simply lift one side and roll or drag it to fresh ground daily or every few days.

A common question that arises with this sort of setup is whether your shelter needs a floor. Chicken tractors are designed without one so birds can forage directly on the ground. However, if predators like snakes or rats are an issue, you can line the bottom with wire mesh to keep unwanted guests out while allowing scratching.

Speaking of predators, your design should include secure latches, hardware cloth (not chicken wire), and a snug fit where the panels meet the ground to keep out unwanted guests. For egg-laying hens, add nests inside the tractor, ideally raised and accessible for easy egg collection.

When it comes to materials, wood chicken tractors are sturdy but heavy and prone to rot; plastic ones are lightweight and weather-resistant but may warp under heat; and metal frames are durable yet require rust protection. You can use a mix of these materials for best results. For example, wood for structure and metal or plastic for roofing.

Covers are essential for shade and rain protection. Use corrugated plastic, tarp, or metal sheets for the roof, leaving some ventilation open. For fasteners, choose rust-resistant screws and bolts to ensure longevity and easy maintenance.

Finally, moving the shelter regularly is key to keeping your pasture healthy and your chickens clean. Relocate daily or every couple of days to prevent manure buildup and give the grass time to recover.

Rotational Grazing

Rotational grazing is one of the smartest, most sustainable ways to keep chickens whether you live in the suburbs or own a small homestead in the countryside. Instead of letting your flock stay in one run (where grass quickly turns to dust), you move them between different patches of land in a planned cycle. This gives each area time to rest and regrow while the chickens enjoy fresh forage, insects, and space to roam. It's a system that mimics nature's balance where nothing goes to waste, and everything has time to renew.

Even for urban chicken keepers, rotational grazing is completely doable. By dividing your yard into smaller sections or "mini pastures," you can rotate your flock through each zone every few days or weeks. While one section recovers, the chickens fertilize the next with their nutrient-rich manure, aerate the soil with their scratching, and naturally manage pests and weeds.

This idea builds directly on chicken tractors, which we explored earlier. A chicken tractor makes rotational grazing simple. Move it regularly, and your chickens get fresh ground to explore while the soil beneath receives a boost of nitrogen and organic matter. Over time, this cycle transforms even poor soil into rich, dark earth perfect for gardening.

For those without tractors, movable fencing is an excellent alternative. Lightweight electric netting or portable mesh panels can be arranged to create temporary paddocks. This allows you to manage where the flock grazes while keeping them safe from predators and out of sensitive areas like vegetable beds.

Rotational grazing also fits beautifully into no-till gardening. Chickens are natural soil workers. They dig, weed, and mix organic matter effortlessly. Before planting, you can let them prepare a garden bed for you: they'll eat pests, clear weeds, and fertilize as they go. Afterward, simply level the soil and plant.

The first grazed patch, after you embrace rotational grazing for your hens, may look messy, but soon gives way to greener and thicker grass, making it all worth it. This is nature's way of showing how rest, rotation, and regeneration work in harmony.

Of course, this method does have a few challenges. It takes planning, time, and a bit of muscle to move fencing or coops regularly. Smaller yards may need longer rest periods for grass to recover, and rainy weather can turn grazed sections muddy. But once you find your rhythm, the process becomes second nature.

The benefits of this practice easily outweigh the effort. You'll have healthier chickens, richer soil, fewer pests, and nutrient-packed eggs. It's a simple yet powerful way to create a regenerative cycle where your flock and your land support each other naturally.

Keeping Records: Why It Matters

I used to think record-keeping was for "serious" farmers, people with spreadsheets, clipboards, and probably flannel shirts. Me? I had five chickens. Surely I could remember who laid what and when.

Spoiler: I couldn't.

By year two, I had no idea which hens were my best layers, which ones had health issues, or how much I was actually spending on feed. When one hen got sick, I couldn't remember when I'd last dewormed the flock. When egg production dropped, I had no baseline to compare it to.

That's when I started keeping records. Just simple notes in a notebook. And honestly? It changed everything.

Tracking your flock doesn't have to be complicated or time-consuming. But even basic records can help you:

- **Identify problems early** (sudden drops in egg production, recurring illnesses)
- **Track expenses and income** (especially if you sell eggs)
- **Improve breeding decisions** (if you hatch chicks)
- **Monitor individual hen health and productivity**
- **Make informed decisions** (which hens to keep, which to rehome, when to add new birds)

Let's break down what to track and how to keep it simple.

What to Track

You don't need to record every cluck and peck. Focus on the essentials:

1. Flock Inventory

What to Record:
- Hen's name or ID (leg bands work great)
- Breed
- Hatch date or age when acquired
- Date acquired
- Source (hatchery, breeder, feed store, etc.)
- Physical description (color, markings, unique features)

Why It Matters:

Knowing each bird's age helps you predict laying patterns and plan for replacements. Breed info helps you make decisions about future flock composition.

Example:

- Name: Dorothy
- Breed: Rhode Island Red
- Hatch Date: 3/15/2020
- Acquired: 5/1/2020
- Source: Local hatchery
- Notes: Bossy, great layer, curious

2. Egg Production

What to Record:

- Daily egg count (total or per hen if you can tell)
- Weekly or monthly totals
- Any abnormalities (soft shells, double yolks, blood spots)

Why It Matters:

Tracking production helps you spot patterns, identify your best layers, and notice drops that might signal illness, stress, or seasonal changes.

Example:

Date	Total Eggs	Notes
11/1	4	Normal
11/2	3	Dorothy didn't lay
11/3	5	One double yolk (Bella)

Pro Tip: You don't have to record daily. Weekly totals work fine for casual keepers.

3. Health & Medical Records

What to Record:

- Symptoms observed (lethargy, sneezing, limping, etc.)
- Treatments given (medications, natural remedies, dosages)
- Vet visits and diagnoses
- Deworming dates
- Vaccinations (if applicable)
- Parasites treated (mites, lice, worms)
- Outcome (recovered, ongoing, deceased)

Why It Matters:

Health records help you identify recurring issues, track treatment effectiveness, and provide critical info to your vet. They also remind you when it's time for routine care (like deworming).

Example:

Date	Hen	Symptoms	Treatment	Outcome
10/15	Dorothy	Sneezing, nasal discharge	Oregano oil in water (5 days)	Resolved
10/20	Entire flock	Routine deworming	Fenbendazole (3 days)	N/A

4. Expenses & Income

What to Record:

- Feed purchases (date, amount, cost)
- Bedding, supplements, treats
- Coop repairs or upgrades
- Medical expenses (vet bills, medications)
- Equipment purchases (feeders, waterers, etc.)
- Income from egg sales, bird sales, or manure

Why It Matters:

Knowing your costs helps you budget, determine if you're breaking even (if selling), and make informed decisions about flock size.

Example:

Date	Category	Item	Cost
10/5	Feed	50 lb layer feed	$18
10/12	Bedding	Pine shavings	$7
10/20	Income	Egg sales (2 dozen)	+$12

Pro Tip: Keep receipts in an envelope or use a simple spreadsheet. Review quarterly to see if you're on track.

5. Breeding & Hatching (If Applicable)

What to Record:

- Breeding pairs (rooster + hen)
- Egg set date
- Candling results (Day 7, Day 14)
- Hatch date
- Hatch rate (how many eggs hatched vs. set)
- Chick health and development

Why It Matters:

If you breed chickens, records help you identify your best breeding stock, track hatch success, and avoid inbreeding.

Example:

Set Date	Rooster	Hen	Eggs Set	Hatched	Hatch Rate	Notes
3/1	Rusty (RIR)	Dorothy (RIR)	12	9	0.75	3 clear eggs

How to Keep Records (Without Losing Your Mind)

1. The Notebook Method

Keep a small notebook in or near the coop. Jot down daily observations, egg counts, and health notes. At the end of the week or month, transfer key info to a summary page or leave it as-is.

Pros:
- Simple, low-tech
- Always accessible
- No learning curve

Cons:
- Can get messy or disorganized
- Hard to search or analyze data
- Risk of losing the notebook

2. The Spreadsheet Method

Create tabs for Inventory, Egg Production, Health, Expenses, etc. Update weekly or as needed. Use formulas to calculate totals, averages, or trends.

Pros:
- Organized and searchable
- Easy to analyze trends
- Can share with vets or other keepers

Cons:
- Requires computer/device access
- Steeper learning curve
- Can feel overwhelming at first

Pro Tip: Use Google Sheets so you can update from your phone in the coop!

3. The App Method

Download a chicken-keeping app like:
- **Flock Tracker** (iOS, Android)
- **Chicken Manager** (Android)
- **Chicken Kit** (iOS)
- **My Chicken Flock** (iOS, Android)

These apps let you track inventory, eggs, health, expenses, and more, all in one place.

Pros:
- Convenient (always with you)
- Built-in features (reminders, charts, photos)
- Cloud backup

Cons:
- Requires smartphone
- Some apps have limited free features
- Learning curve for new apps

What I Actually Do

Here's my system, which takes about 5 minutes a day:
- **Morning:** I grab my phone and open Google Sheets. I note the date, egg count, and any observations (e.g., "Dorothy looks off").
- **Weekly:** I tally expenses (feed, bedding) and income (egg sales).
- **Monthly:** I review trends; are eggs dropping? Is one hen consistently underperforming? Do I need to budget for a vet visit?
- **Annually:** I assess each hen's productivity and decide if anyone needs to be retired or rehomed.

It's not perfect, but it works. And that's the goal, find a system that fits your life.

Sample Record Templates

Basic Egg Production Log:

Week Of	Mon	Tue	Wed	Thu	Fri	Sat	Sun	Total
11/1	4	5	4	3	5	4	5	30

Simple Expense Tracker:

Month	Feed	Bedding	Medical	Other	Total
October	$40	$15	$0	$10	$65

Health Log:

Date	Hen	Issue	Treatment	Follow-Up
10/15	Dorothy	Sneezing	Oregano oil	Resolved 10/20

When Records Save the Day

Last winter, I noticed my egg count dropped from 25-30 per week to 10-12. I pulled up my spreadsheet and realized:

1. The drop started exactly when daylight hours shortened (expected).
2. But it was *more* dramatic than previous winters.
3. Two specific hens hadn't laid in weeks.

I checked those hens and found they were both dealing with scaly leg mites (which I'd missed during daily checks). I treated them, and production slowly recovered.

Without records, I might have assumed it was just winter and missed a real problem.

The Bottom Line

You don't need to be a data nerd to keep records. Even basic notes: egg counts, health issues, major expenses, can help you be a better chicken keeper.

Start simple. Pick one or two things to track (like egg production and expenses) and build from there. Use whatever method feels easiest: notebook, app, or spreadsheet. The best record-keeping system is the one you'll actually use.

And remember: the goal is having enough info to make smart decisions and keep your flock healthy and happy.

Record-Keeping Starter Checklist

☐ Choose your tracking method (notebook, spreadsheet, app, or hybrid)
☐ Create a flock inventory (names, breeds, ages)
☐ Start tracking daily or weekly egg production

☐ Note health issues and treatments
☐ Record expenses (feed, bedding, supplies)
☐ Set reminders for routine care (deworming, seasonal checks)
☐ Review records monthly to spot trends
☐ Adjust system as needed, keep it simple!

Good records help you understand your flock, catch problems early, and keep your hens thriving year-round. And when your girls are healthy and laying well, you're rewarded with the best part of chicken keeping: fresh, beautiful eggs.

Now let's put those eggs to good use and explore some simple, delicious recipes straight from your coop to your kitchen.

The Takeaway

By now, you've seen how deeply interconnected chickens are with nature. From natural pest control and soil enrichment to compost creation and rotational grazing, your flock isn't just producing eggs, they're working with you to restore the land. In this chapter, we explored how chickens can cap your pest problems through CAP'M, how to set up a chicken-friendly garden, and how to use chicken tractors and rotational grazing to create a regenerative system that keeps both your birds and your soil thriving.

We also learned how chicken manure becomes gold for the garden, how brooders and roosters shape flock dynamics, and how thoughtful flock management can transform ordinary backyard chickens into the beating heart of an eco-friendly homestead. With the techniques provided in this book, you'll soon have a healthy, happy flock and more eggs then you know what to do with. The Bonus Chapter is full of recipes so not a single one of those hard earned eggs goes to waste.

Conclusion

A Note from Me to You

When I think back to my first morning collecting eggs, I still remember the excitement, the soft rustle of straw, the curious gaze of a hen, and that small, perfect egg resting warm in my palm. It wasn't just food; it was proof that I could grow something wonderful in my backyard with just a bit of care and consistency.

That's really what this journey comes down to. Raising chickens reconnects us with something real. In this fast-paced world, it gives us a reason to slow down. It teaches patience, gentleness, and gratitude. Every day becomes a reminder that life can be simple and still feel full.

You've learned how to keep your flock healthy, protect them from harm, and read their little signals of joy or distress. You've seen that your birds are more than just egg-layers. They're companions, comedians, and teachers. With each passing season, your coop will change, your hens will grow older, and so will your confidence.

Someday soon, you'll step outside at dawn, coffee in hand, and smile at the sounds of your contented flock. The air will feel a little fresher, the earth a little more generous. And you'll realize that your backyard has become a small, thriving world of its own: a reminder that sustainability begins with the small, thoughtful things we choose to do each day.

So, as you turn the last page of this book, picture yourself stepping into the yard one morning to find that first warm egg resting in the nesting box. It's a small thing, but it marks the beginning of something meaningful: a rhythm, a relationship, a way of living a little closer to the earth. May your days ahead be filled with clucks, feathers, laughter, and the deep satisfaction of tending a life that gives back to you in its own simple way.

- Sophie

Thanks for Reading, Please Leave a Review!

I would be *incredibly appreciative* if you could rate my book or leave a review on **Amazon**.

Just scan this QR code with your phone, or visit the https://chicken.sophiemckay.com link to land directly on the book's Amazon review page.

Your review not only helps me create better books, but also helps more fellow gardener experience success in the garden and put healthy food on their family's table.

Thank you!

Sophie

Bonus Chapter 1.

Egg Recipes

Once those eggs start rolling in, you'll find yourself looking for creative ways to put them to good use. And what better way is there to use eggs than to eat them? But let's be honest, seeing the same kind of omelette or hard boiled egg at the table every morning is bound to kill our appetite. So here are some fantastic recipes for a delicious and eggciting breakfast every morning.

1. Shakshuka

A North African and Middle Eastern classic, this dish includes eggs poached in a spicy tomato and pepper sauce.

Ingredients
- 2 tbsp olive oil
- 1 onion, diced
- 1 red bell pepper, diced
- 3 cloves garlic, minced
- 1 tsp ground cumin
- 1 tsp paprika
- ¼ tsp chili flakes (optional)
- 1 can (400 g) crushed tomatoes
- Salt and pepper, to taste
- 4–6 eggs
- Fresh parsley or cilantro, chopped (for garnish)

Instructions

1. Heat olive oil in a large skillet over medium heat.
2. Add onion and bell pepper; cook 5–7 minutes until softened.
3. Stir in garlic, cumin, paprika, and chili flakes; cook 1 minute until fragrant.
4. Pour in tomatoes, season with salt and pepper, and simmer for 10–12 minutes until thickened.
5. Make small wells in the sauce and crack eggs into each.
6. Cover and cook for 5–8 minutes, until eggs are set to your liking.
7. Garnish with herbs and serve with crusty bread.

2. Cheesy Spinach & Artichoke Baked Eggs

Creamy, cheesy, and hearty, this is the perfect egg recipe for brunch or an easy dinner.

Ingredients

- 1 tbsp butter
- 1 cup spinach, chopped
- ½ cup canned artichoke hearts, chopped
- ¼ cup cream cheese
- ¼ cup shredded mozzarella
- 2 tbsp grated Parmesan
- 4 eggs
- Salt and pepper, to taste

Instructions

1. Preheat the oven to 375°F (190°C).
2. Melt butter in an oven-safe skillet. Add spinach and artichokes; cook 2–3 minutes until wilted.

3. Stir in cream cheese, mozzarella, and Parmesan until creamy.
4. Make four small wells and crack one egg into each.
5. Season with salt and pepper.
6. Transfer to the oven and bake for 10–12 minutes, until eggs are just set.
7. Serve warm with toast or pita bread.

3. Eggs in Purgatory

An Italian comfort dish, this dish features eggs simmered in garlicky tomato sauce with a spicy kick.

Ingredients:
- 2 tbsp olive oil
- 3 cloves garlic, sliced
- ½ tsp red pepper flakes
- 1 can (400 g) crushed tomatoes
- Salt and pepper, to taste
- 4 eggs
- Fresh basil, for garnish
- Crusty bread, for serving

Instructions:
1. Heat olive oil in a skillet over medium heat.
2. Add garlic and red pepper flakes; cook until fragrant, about 1 minute.
3. Stir in tomatoes, season with salt and pepper, and simmer for 10 minutes.
4. Make small wells in the sauce and crack in the eggs.
5. Cover and cook until eggs are done to your liking, about 5 minutes.
6. Garnish with basil and serve with bread.

4. Ottolenghi's Braised Eggs with Leek and Za'atar

A fragrant Middle Eastern-style egg dish with rich leeks, herbs, and za'atar spice.

Ingredients:

- 3 tbsp olive oil
- 2 large leeks, sliced
- 1 garlic clove, minced
- 1 tsp cumin seeds
- 1 tbsp za'atar
- ½ tsp chili flakes
- 200 g baby spinach
- 4–6 eggs
- Salt and pepper, to taste
- Yogurt or labneh (optional, for serving)

Instructions:

1. Heat olive oil in a large skillet over medium heat.
2. Add leeks and cook gently for 10 minutes until soft.
3. Stir in garlic, cumin, za'atar, and chili flakes; cook 1–2 minutes.
4. Add spinach and cook until wilted.
5. Make small wells and crack in the eggs.
6. Cover and cook until eggs are set but still soft, 4–6 minutes.
7. Season with salt, pepper, and an extra sprinkle of za'atar. Serve with yogurt and bread.

5. Creamy Bacon and Egg Pasta with Bacon-Fried Eggs

A rich, comforting twist on carbonara. This creamy pasta with crispy bacon and fried eggs is sure to win hearts at the breakfast table.

Ingredients:
- 200 g spaghetti or fettuccine
- 6 slices bacon, chopped
- 2 eggs + 2 for frying
- ½ cup grated Parmesan
- ½ cup cream
- Salt and pepper, to taste

Instructions:
1. Cook pasta in salted boiling water until al dente; reserve ½ cup pasta water.
2. In a skillet, cook bacon until crisp. Remove half for topping.
3. In a bowl, whisk 2 eggs, Parmesan, and cream.
4. Add cooked pasta to the skillet with bacon; remove from heat.
5. Quickly stir in the egg mixture, tossing well. Add a little pasta water to make it creamy.
6. In another pan, fry the remaining 2 eggs in bacon fat until the edges are crisp.
7. Serve the creamy pasta topped with bacon pieces and fried eggs.

Bonus Chapter 2.

Coop Plans

If you want those eggs coming, you've got to give your hens a comfortable place to stay. Here are some budget-friendly coop plans that are perfect for beginners.

1. Walk-In Chicken Coop Plan (20′ × 9′)

Size & Capacity: 20′ long × 9′ wide; suitable for approximately 12 chickens.

Materials Required:

- Pressure-treated lumber for floor joists / rim: e.g., 2×8 or 2×10 treated, lengths per plan
- Standard lumber for wall framing (e.g., 2×4 studs, headers)
- Plywood sheathing for floor, walls, roof
- Siding boards (exterior grade)
- Roof sheathing + roofing material (asphalt shingles or metal), drip edge & flashing
- Hardware cloth/galvanized wire mesh for run & ventilation openings
- Doors (walk-in size), hinges, latches, handles
- Window(s) for light & ventilation
- Fasteners: wood screws (e.g., 3", 2½"), galvanized nails, corner braces, metal brackets
- Roof underlayment / building felt as needed
- Paint / stain / sealant for exterior
- Nesting boxes, roost bars, ramp for chickens

Instructions:

1. **Site preparation & foundation:** Clear and level ground. Lay concrete blocks or pavers or build a treated timber frame to sit the floor joists on.
2. **Floor framing:** Cut joists per plan (for example six boards cut to 5'-1" in one step) and assemble rim joists and interior joists.
3. **Floor sheathing:** Secure plywood over the framing, ensuring it is level and anchored properly.
4. **Wall framing:** Build front wall (with door opening) to plan: for example four boards at 8'-3", two at 5'-10 ½" for studs, etc. Build side walls and back walls similarly. Raise

and secure walls to the floor frame, ensure plumb and square.

5. **Roof framing:** Install rafters or roof joists per plan. Add sheathing (plywood), underlayment, and roofing material (shingles/metal). Install drip edge and any flashing at intersections.
6. **Exterior sheathing and siding:** Attach plywood sheathing to walls, then siding boards. Install windows and doors. Seal all joints, apply paint or stain.
7. **Run/enclosure & predator proofing:** Frame the run area (9' width × 14-15' length per plan) with framing lumber and cover with hardware cloth. Bury or footer the mesh into the ground for predator prevention. Add access doors/gates.
8. **Interior fittings:** Install nesting boxes (for 12 chickens go for 8 nesting boxes). Install roost bars elevated above the nesting boxes, install ramps from run to coop floor if required.
9. **Ventilation & finishing touches:** Install vents high on walls or just under eaves, ensure airflow. Add feeders/waterers, bedding, check doors/hatches for weatherproofing.
10. **Inspection & move-in:** Walk around, check all fasteners, ensure no sharp edges or openings. Clean interior, add bedding, introduce your flock. Monitor for the first few days for any issues.

2. Extra-Large Chicken Coop with Run (9′ × 42′)

Size & Capacity: 9′ wide × 42′ long, capacity up to 20 chickens.

Materials:

- Treated lumber for foundation/ground contact
- Lumber for framing large run (long spans) and coop section
- Plywood sheathing for floor, walls, roof
- Siding boards, roofing materials
- Hardware cloth for extensive run area
- Doors/gates for both coop and run access
- Ventilation components (lots of ventilation due to long run)
- Fasteners, brackets, corner braces
- Paint/stain/sealant
- Nesting boxes (6 or more, as plan suggests)
- Roosts, ladder/ramps possibly

Instructions:

1. **Site prep & foundation:** Clear and level a very large area (9′ × 42′) with good drainage. Lay treated base or concrete blocks/pavers spaced along the length to support the structure.
2. **Floor framing:** Construct floor for coop section and maybe elevated section if needed; the long run may sit on ground or low frame depending on design.
3. **Wall framing for coop section:** Build the coop inlet (width likely 9′, length maybe ~5′-6′ depending). Frame height per plan. Raise walls.
4. **Roof framing and covering for coop:** Build roof structure (gable or lean-to) over coop section and possibly a covered run area. Sheath and cover with roofing material.
5. **Run structure:** Build the long run frame (42′ length) using posts/rafters/trusses as needed. Cover all sides and top with hardware cloth to protect from aerial and ground predators. Ensure the mesh is secured and buried or footed.
6. **Sheathing and siding for coop:** Attach plywood to walls and siding boards. Install windows/vent openings. Paint or stain the exterior.
7. **Doors and access:** Install full-sized walk-in door for coop section and one or more access doors in the run for cleaning and maintenance.
8. **Interior fittings:** Install nesting boxes (6 or more), roost bars elevated above the floor, ramps if necessary. Provide feeders and waterers spread across the run.
9. **Ventilation & predator proofing:** Due to size, ensure multiple vents high and low in the coop section. Check run

for predator ingress (digging, climbing). Install locks/latches that predators (raccoons, foxes) cannot open.

10. **Final check & introduction:** Walk the entire length to inspect for gaps or weak spots. Clean inside the coop, add bedding, move chickens in, and monitor behavior and coop integrity over the first week.

3. DIY Chicken Tractor Plans (4' × 8')

Size & Capacity: 4' wide × 8' long, suitable for a small flock or garden rotational use.

Materials Required:
- Treated lumber or decay-resistant lumber for base frame
- Lumber for short walls and roof framing
- Plywood or light sheathing for coop compartment
- Hardware cloth/galvanized mesh for run area all around
- Wheels or skids (if tractor style) or handles for moving
- Roofing material (metal sheet or shingles)
- Doors/hatches and fasteners

- Paint/stain/sealant
- Nesting box and roost bar (for a smaller number of chickens)

Instructions:

1. **Frame base:** Construct a 4′ × 8′ rectangle base. If mobile, attach wheels or sliders/skids at corners so you can move it.
2. **Floor sheathing:** If there is an enclosed coop portion, sheath that floor with plywood.
3. **Build coop section:** On one end of the base, build short walls for the coop (for roosting/nesting) and a roof. Sheath and cover.
4. **Frame the run area:** The open portion (or mostly open) allows chickens to roam. Frame the perimeter and overhead (if predator protection overhead is needed) and attach hardware cloth. Bury mesh at edges or attach footplates to prevent digging.
5. **Roofing & weatherproofing:** Cover coop section roof with roofing material and ensure run overhead has coverage or shielding as desired.
6. **Doors & mobility finishing:** Install coop door and run access door/latch. Attach handles, ensure movement is smooth, check for tipping.
7. **Interior fittings:** Install a small nesting box and roost bar sufficient for the number of birds (e.g., 4-8).
8. **Finishing touches:** Paint or seal the exterior lumber, remove sharp edges from wire mesh, add feeders/waterers inside run or coop.

9. **Placement & rotation strategy:** Place on fresh grass/soil, move it every few days or week to fresh ground. Monitor ground conditions and adjust.

4. Hen's Paradise Gable Coop (4' × 10')

Size & Capacity: Approx. 4' wide × 10' long – a smaller gable-roofed coop.

Materials Required:
- Treated lumber for base/floor/joists
- Lumber for gable roof framing and walls
- Plywood sheathing for floor, walls, roof
- Exterior siding boards
- Roof covering (metal or shingles)
- Hardware cloth for windows/ventilation and run if included
- Doors, hinges, latches
- Fasteners, brackets, corner braces

- Paint/stain/sealant
- Nesting boxes and roost bars (for a smaller flock)

Instructions:

1. **Site and base:** Clear and level site. Build the base frame using treated lumber.
2. **Floor sheathing:** Attach plywood to the base.
3. **Wall framing:** Build front/back and side walls to 4' wide × 10' long. Include window/door openings. Raise walls and secure.
4. **Gable roof framing:** Construct gable trusses or rafters at each end and ridge. Sheath with plywood, attach roofing material.
5. **Exterior finishing:** Attach siding boards, install windows/vents, door. Apply paint/stain/sealant.
6. **Run/ventilation:** If there's an attached run, frame and mesh it with hardware cloth. If not, at least add ventilation openings with mesh covers.
7. **Interior fittings:** Install nesting boxes (maybe 2-3 for a small flock), roosting bars.
8. **Finishing & cleanup:** Seal edges, ensure no exposed mesh sharp edges, add feeders/waterers.
9. **Introduce birds & observe:** Place chickens, ensure they use roosts and nesting boxes, monitor for comfort and safety.

As you begin building a comfortable coop for your hens, here are some points to bear in mind.

- **Predator proofing is non-negotiable**: Bury hardware cloth 6–12" into ground, extend outwards at base, use secure latches that raccoons can't open.

- **Ventilation matters**: Regardless of size, fresh air reduces ammonia, moisture build-up and helps chickens stay healthy.
- **Easy cleaning**: Try to include access doors, removable trays, or easy-to-reach interior so you can clean properly.
- **Floor space guidelines**: As a rule of thumb many chicken keepers recommend 4 sq ft per bird inside the coop and 10 sq ft or more in the run, though it depends on breed and climate.
- **Use treated lumber / rot-resistant materials** especially at ground or contact with soil.
- **Adapt to your climate**: If you're in a colder region, you may want more insulation, fewer large windows, etc. For hot climates emphasize shade and air flow.
- **Check local zoning/building codes**: Especially for larger builds (9×42 or 20×9) you may need permits or adhere to setbacks.
- **Measure twice, cut once**: Use proper squares, levels and ensure your framing is square, especially in long spans (like the 42′ length).
- **Safety first**: For moving builds (tractor style) ensure wheels/skids are robust, materials anchored, and the coop won't tip or collapse under movement.

Appendix A

Troubleshooting Common Problems / Quick Reference

Quick Fixes for Common Flock Problems

Chickens are generally easy to care for, but every keeper eventually faces a head-scratching moment: Why won't they lay? Why are they fighting? Why is that one acting so *weird*?

Here's a quick-reference guide to the most common problems and their solutions. Bookmark this page—you'll thank me later.

Problem 1: Suddenly No Eggs (Or Fewer Eggs)

Possible Causes:

- Molting (fall/late summer)
- Short daylight hours (winter)
- Stress (predator scare, new birds, coop changes)
- Broodiness
- Age (hens slow down after 2-3 years)
- Poor nutrition
- Illness or parasites
- Egg-eating or hidden nest

Quick Fixes:

☐ Check for molting (feather loss, pin feathers)
☐ Add supplemental lighting if winter (14-16 hours total light)
☐ Reduce stressors (minimize changes, check for predators)
☐ Break broodiness (cooling box, remove nesting access)
☐ Boost protein (18-20% during molt)
☐ Ensure calcium access (oyster shell)
☐ Deworm and check for mites/lice
☐ Search for hidden nests
☐ Collect eggs 2x daily to prevent egg-eating

Problem 2: Aggressive Hen or Rooster

Possible Causes:

- Establishing pecking order (normal)
- Overcrowding
- Resource guarding (food, nest, roost)
- Breed temperament (some breeds are naturally feistier)
- Hormonal behavior (especially roosters)
- Previous trauma or poor socialization

Quick Fixes:

☐ Ensure adequate space (4 sqft per bird minimum)
☐ Add multiple feeders/waterers
☐ Provide hiding spots and visual barriers
☐ Remove bully temporarily (3-5 days) to reset pecking order
☐ Use distractions (hanging treats, scratch grains)
☐ Consider rehoming if aggression is extreme/dangerous
☐ For roosters: assess hen-to-rooster ratio (1 rooster per 8-10 hens ideal)

Problem 3: Feather Loss (Not Molting)

Possible Causes:

- Mites or lice infestation
- Feather pecking/bullying
- Over-mating by rooster
- Poor nutrition (protein deficiency)
- Stress or boredom
- Vent prolapse or vent pecking

Quick Fixes:

☐ Inspect for parasites (check under wings, around vent, on roosts at night)
☐ Treat for mites/lice (diatomaceous earth, poultry dust, natural sprays)
☐ Add protein to diet (mealworms, game bird feed, scrambled eggs)
☐ Provide dust bath area with sand and wood ash
☐ Use chicken saddles on hens being over-mated
☐ Add enrichment (hanging vegetables, perches, foraging opportunities)
☐ Separate bullied birds temporarily
☐ Check rooster-to-hen ratio (too few hens = over-mating)

Problem 4: Broody Hen Won't Leave Nest

Possible Causes:

- Natural hormonal cycle (instinct to hatch eggs)
- Certain breeds are more prone (Silkies, Cochins, Orpingtons)
- Accumulation of eggs in nest box

Quick Fixes (If You DON'T Want Chicks):

☐ Remove her from nest several times daily
☐ Block access to nesting boxes
☐ Place her in a wire-bottom cage with food/water but no bedding (3-5 days)
☐ Ensure good ventilation to cool her underside
☐ Collect eggs frequently so they don't accumulate
☐ Provide distractions and encourage activity

If You DO Want Chicks:

☐ Move her to a separate broody box
☐ Provide 10-12 fertile eggs
☐ Ensure she has food, water, and quiet space
☐ Let her sit for 21 days undisturbed

Problem 5: Chickens Won't Go Into Coop at Night

Possible Causes:

- New birds unfamiliar with coop
- Predators or pests inside coop (mites, snakes, rats)
- Poor coop conditions (too hot, smelly, cramped)
- Roosting bars uncomfortable or too high
- Prefer outdoor roosts (trees, fences)

Quick Fixes:

☐ Manually place birds on roost at dusk for 3-7 nights (they'll learn)
☐ Check for mites on roosts (inspect at night with flashlight)
☐ Clean coop thoroughly, replace bedding
☐ Improve ventilation and reduce ammonia smell
☐ Lower roosting bars or add a ramp
☐ Block access to alternative roosting spots outside

☐ Add a light inside coop at dusk to attract them in
☐ Use treats to lure them inside before dark

Problem 6: Smelly Coop

Possible Causes:

- Dirty bedding (ammonia buildup from droppings)
- Poor ventilation
- Wet bedding (leaky waterer, rain, humidity)
- Overcrowding
- Infrequent cleaning

Quick Fixes:

☐ Clean coop immediately and replace all bedding
☐ Improve ventilation (add vents near roof, not at bird level)
☐ Use deep litter method (add fresh bedding weekly, deep clean quarterly)
☐ Fix leaky waterers or move them outside
☐ Add diatomaceous earth or barn lime to bedding (odor control)
☐ Reduce flock size if overcrowded
☐ Use droppings boards under roosts for easier daily cleaning

Problem 7: Bullying or Pecking Order Violence

Possible Causes:

- New bird introductions (disrupts hierarchy)
- Overcrowding
- Boredom or lack of enrichment
- Injured or weak bird attracting aggression
- Insufficient resources (food, water, nests, roosts)

Quick Fixes:

☐ Separate injured bird immediately (blood attracts more pecking)
☐ Treat wounds with Blu-Kote or antiseptic
☐ Increase space (expand run or reduce flock size)
☐ Add multiple feeders/waterers
☐ Provide distractions (hanging cabbage, scratch grains, toys)
☐ Use pinless peepers on aggressive birds (reduces pecking accuracy)
☐ Remove bully temporarily to reset pecking order
☐ Add hiding spots and visual barriers

Problem 8: Chickens Eating Eggs

Possible Causes:

- Accidental breakage (thin shells, overcrowding)
- Curiosity (hen discovers eggs taste good)
- Nutritional deficiency (protein or calcium)
- Boredom

Quick Fixes:

☐ Collect eggs 2-3 times daily
☐ Add more nesting boxes (1 per 3-4 hens)
☐ Use deeper bedding in nests (2-3 inches)
☐ Provide extra calcium (oyster shell)
☐ Boost protein (18-20% feed, mealworms)
☐ Use fake eggs or golf balls in nests (hens get frustrated and stop)
☐ Use roll-away nesting boxes (eggs roll out of reach)
☐ Fill empty eggshells with mustard or hot sauce as deterrent
☐ Add enrichment to reduce boredom

Problem 9: Soft or Thin-Shelled Eggs

Possible Causes:

- Calcium deficiency
- Vitamin D deficiency
- Stress or illness
- Heat stress (reduces calcium absorption)
- Young pullet just starting to lay
- Older hen nearing end of laying years

Quick Fixes:

☐ Provide free-choice oyster shell or crushed eggshells
☐ Ensure flock gets adequate sunlight (or add Vitamin D supplement)
☐ Switch to high-quality layer feed with proper calcium (3.5-4%)
☐ Reduce stressors (predators, overcrowding, sudden changes)
☐ Provide shade and cool water during hot weather
☐ Monitor young pullets (soft shells often resolve naturally)
☐ Accept reduced quality in older hens

Problem 10: Chickens Won't Use Nesting Boxes

Possible Causes:

- Boxes too high, too small, or uncomfortable
- Boxes too bright or exposed
- Preferred hidden spot elsewhere
- Mites or pests in boxes
- New birds unfamiliar with boxes

Quick Fixes:

☐ Make boxes dark and private (add curtains, move to quiet corner)
☐ Ensure boxes are large enough (12" x 12" minimum)
☐ Add 2-3 inches of soft bedding
☐ Place fake eggs or golf balls in boxes (signals "this is where we lay")
☐ Block access to alternate laying spots
☐ Check for mites and treat if needed
☐ Collect eggs daily so boxes don't appear "full"
☐ Place young pullets in boxes at dusk to familiarize them

Problem 11: Chicken Acting Lethargic or "Off"

Possible Causes:

- Illness (respiratory, digestive, reproductive)
- Parasites (worms, mites, lice)
- Egg-binding
- Crop issues (impacted or sour crop)
- Injury
- Heat stress or cold stress
- Poisoning (toxic plants, moldy feed)

Quick Fixes:

☐ Isolate bird immediately in a quiet, comfortable space
☐ Check for obvious injuries (wounds, swelling, limping)
☐ Feel crop (should be empty in morning, full after eating)
☐ Check vent (should be moist, not swollen or prolapsed)
☐ Offer electrolytes in water
☐ Provide favorite treats to encourage eating
☐ Monitor droppings (color, consistency)
☐ Check for respiratory symptoms (sneezing, discharge, labored breathing)

☐ Deworm if not done recently
☐ Consult vet if no improvement in 24-48 hours

Problem 12: Chickens Drinking Excessive Water

Possible Causes:

- Hot weather (normal)
- High-sodium treats (salty scraps)
- Kidney or liver disease
- Diabetes (rare in chickens)
- Egg production (layers need more water)

Quick Fixes:

☐ Monitor in hot weather (increased drinking is normal)
☐ Reduce salty treats (bread, crackers, processed foods)
☐ Ensure water is clean and cool
☐ Monitor droppings (watery droppings may indicate illness)
☐ Consult vet if excessive drinking continues with other symptoms (lethargy, weight loss)

Problem 13: Rooster Crowing Too Much (Noise Complaints)

Possible Causes:

- Natural behavior (roosters crow to establish territory)
- Multiple roosters competing
- Nearby roosters (they "answer" each other)
- Excitement (sunrise, feeding time, new visitors)

Quick Fixes:

☐ Accept that roosters crow (it's what they do!)
☐ Use a rooster collar – check with vet first (reduces volume by restricting airflow slightly)
☐ Keep coop dark until later in morning (delays crowing)
☐ Reduce flock to one rooster
☐ Rehome rooster if noise complaints are serious
☐ Check local ordinances (many cities ban roosters)

Problem 14: Chickens Escaping Run or Yard

Possible Causes:

- Fence too low (chickens can fly 4-6 feet)
- Gaps or holes in fencing
- Motivated foragers seeking better food/space
- Predator pressure (trying to escape danger)

Quick Fixes:

☐ Raise fence height to 6 feet or add netting on top
☐ Clip flight feathers on one wing (makes flying unbalanced)
☐ Check for gaps, holes, or weak spots in fencing
☐ Add visual barriers (tarps, solid panels) to reduce urge to escape
☐ Improve run conditions (more space, enrichment, better food)
☐ Free-range supervised instead of confining unhappy birds

Problem 15: Prolapsed Vent

Possible Causes:

- Straining to lay oversized egg
- Obesity

- Egg-binding
- Chronic straining (constipation, diarrhea)
- Young pullet laying too early
- Nutritional imbalance

Quick Fixes:

☐ **Act immediately—this is serious**
☐ Isolate bird in a quiet, dark space
☐ Gently clean prolapse with warm water
☐ Apply hemorrhoid cream or coconut oil to prolapse
☐ Gently push tissue back inside (wear gloves, be very gentle)
☐ Hold in place for several minutes
☐ Keep hen isolated and quiet for 48 hours
☐ Offer soft foods and reduce stress
☐ **If prolapse recurs or worsens, consult vet immediately**
☐ Prevent by avoiding early laying (don't supplement light for pullets under 18 weeks)

Quick Troubleshooting Flowchart

Is the problem related to...

→ EGGS?

- Not laying → Check daylight, nutrition, stress, age, health
- Weird eggs → See Chapter 4 "Egg Abnormalities Guide"
- Eating eggs → Collect often, add fake eggs, increase protein/calcium

→ BEHAVIOR?

- Aggression → Add space, multiple resources, distractions
- Won't go in coop → Manually train, check for pests, improve conditions
- Broodiness → Break it (cooling box) or let her hatch

→ HEALTH?

- Lethargy → Isolate, check for injury/illness, offer electrolytes, consult vet
- Feather loss → Check for parasites, bullying, over-mating, nutrition
- Respiratory issues → Improve ventilation, add oregano/garlic, consult vet

→ COOP/RUN?

- Smelly → Clean, improve ventilation, fix water leaks
- Escaping → Raise fence, clip wings, improve conditions
- Won't use nests → Make dark/private, add fake eggs, ensure comfort

The Golden Rules of Troubleshooting

1. **Observe first, act second.** Watch your flock's behavior for patterns before jumping to conclusions.

2. **Start with the simplest solution.** Often the fix is easier than you think (more food, cleaner coop, extra space).

3. **One change at a time.** If you change multiple things at once, you won't know what worked.

4. **When in doubt, consult experts.** Join online chicken groups, call your local extension office, or contact a vet.

5. **Trust your instincts.** You know your flock best. If something feels wrong, investigate.

6. **Keep records.** Patterns become obvious when you track egg production, expenses, and health issues over time.

The Bottom Line

Most chicken problems have straightforward solutions. The key is staying calm, observing carefully, and taking action quickly when needed.

Bookmark this section, keep it handy near your coop, and remember: every chicken keeper faces these challenges. You're learning, you're adapting, and you're doing great.

APPENDIX B

Humane Euthanasia Methods

Note to Readers: This appendix contains detailed instructions for at-home euthanasia methods. These techniques are intended for experienced keepers, homesteaders, or those in rural areas without access to veterinary care. The information is provided in a clinical, educational manner to ensure proper technique and minimize suffering.

If you are uncomfortable with these methods, or if you have access to veterinary services, we strongly encourage you to choose professional euthanasia. There is no shame in this choice, it is often the most compassionate option for both you and your hen.

The methods described here are widely accepted as humane when performed correctly. However, they require confidence, proper technique, and emotional readiness. If you have any doubts, please seek help from an experienced mentor, agricultural extension agent, or veterinarian.

General Principles of Humane Euthanasia

Regardless of method, humane euthanasia must meet these criteria:

1. **Instantaneous or near-instantaneous unconsciousness** (the bird feels no pain)
2. **Rapid death** following loss of consciousness
3. **Minimal stress** before the procedure
4. **Performed by someone confident and capable** of executing the technique correctly

What Is NOT Humane:

- Drowning
- Suffocation (without proper CO_2)
- Blunt force trauma (unless part of proper cervical dislocation)
- Freezing
- Starvation
- Any method causing prolonged distress or pain

Method 1: Cervical Dislocation

Overview:
Cervical dislocation involves quickly and forcefully dislocating the bird's neck, which severs the spinal cord and causes instant unconsciousness and death. When done correctly, it is immediate and humane.

Best For:
- Experienced keepers comfortable with hands-on methods
- Rural settings where veterinary care is unavailable
- Emergency situations requiring immediate action

What You Need:
- Confidence and physical strength
- Proper technique (practice or training recommended)
- A quiet, private space

Step-by-Step Instructions:

1. **Prepare yourself emotionally.** Take a deep breath. This is an act of mercy, not cruelty.
2. **Hold the bird calmly.** Place the hen on a flat surface or hold her body firmly against your side with one arm, keeping her wings secured.

3. **Grasp the head.** Using your dominant hand, grasp the hen's head from behind, placing your thumb and forefinger on either side of the skull just behind the jawline.

4. **Apply firm, steady pressure.** In one swift, confident motion:
    - Pull the head **downward** and **backward** (toward the tail)
    - Stretch the neck firmly until you feel the vertebrae separate
    - You may hear or feel a "pop" as the spine dislocates

5. **Hold for 10-15 seconds.** Keep the head extended to ensure the dislocation is complete.

6. **Confirm death.** The bird's eyes will glaze, breathing will stop immediately, and the body may have reflexive twitching (this is normal and does not indicate pain).

Important Notes:

- The motion must be **swift and decisive**. Hesitation prolongs suffering.
- If you're unsure of your ability to perform this correctly, seek help or choose another method.
- Watch instructional videos or ask an experienced person to guide you the first time.
- This method is **not** suitable for larger birds (turkeys, geese) unless you have significant strength and experience.

Pros:

- Instantaneous when done correctly
- No equipment needed
- Can be performed anywhere

Cons:

- Emotionally difficult
- Requires physical strength and confidence

- Risk of improper technique if inexperienced

Method 2: Carbon Dioxide (CO_2) Chamber

Overview:
CO_2 euthanasia involves placing the bird in a sealed container with controlled carbon dioxide flow, which causes rapid unconsciousness followed by death. It is considered one of the most humane methods and is widely used in research and veterinary settings.

Best For:

- Keepers seeking a gentler, hands-off method
- Those uncomfortable with physical dispatch methods
- Multiple birds requiring euthanasia

What You Need:

- CO_2 gas tank (available at welding supply stores)
- CO_2 regulator with tubing
- Well-sealed container (cooler, plastic bin with tight-fitting lid)
- Drill or knife to create a small hole in the lid for tubing

Step-by-Step Instructions:

1. **Set up the chamber.** Drill a small hole in the lid of your container, just large enough for the CO_2 tubing to fit snugly.
2. **Insert tubing.** Thread the CO_2 tubing through the hole so it reaches the bottom of the container.
3. **Place the hen inside.** Gently place your hen in the container. You can add a soft towel for comfort.
4. **Secure the lid.** Close the lid tightly to create a seal.
5. **Introduce CO_2 slowly.** Turn on the gas and adjust the regulator to a slow, steady flow (about 10-20%

displacement per minute). Rapid introduction can cause panic.

6. **Observe.** Within 1-2 minutes, the hen will lose consciousness (she may close her eyes, become still, or gently settle). Continue the flow for at least 5 more minutes to ensure death.

7. **Confirm death.** Open the container and check for breathing, heartbeat, and eye reflexes. If unsure, continue CO_2 flow for another 5 minutes.

Important Notes:

- **Do NOT use car exhaust, propane, or other gases.** Only pure CO_2 is humane—other gases cause suffering.
- The bird may vocalize briefly or flutter as she loses consciousness. This is reflexive and does not indicate pain.
- CO_2 is heavier than air, so it will displace oxygen from the bottom up.

Pros:

- Painless and peaceful
- No direct physical handling required
- Suitable for multiple birds

Cons:

- Requires CO_2 tank and setup (initial cost ~$100-$200)
- Can be emotionally difficult to watch
- Must be done correctly to avoid distress

Method 3: Decapitation

Overview:

Decapitation involves quickly severing the head with a sharp tool (hatchet, very sharp knife, or specialized poultry killing cone). When performed correctly with a sharp blade, it causes instantaneous loss of brain function and death.

Best For:

- Homesteaders and farmers comfortable with traditional methods
- Situations requiring quick dispatch
- Keepers with proper tools and experience

What You Need:

- **Very sharp** hatchet, cleaver, or knife (dull blades cause suffering)
- Killing cone (optional but recommended—holds the bird securely)
- Chopping block or sturdy surface
- Confidence and steady hand

Step-by-Step Instructions:

Using a Killing Cone:

1. **Secure the hen.** Place her head-first into a killing cone (a cone-shaped restraint that holds the body while the head protrudes through the bottom).

2. **Extend the neck.** Gently pull the head down so the neck is fully extended.

3. **Sever quickly.** Using a very sharp blade, make one swift, decisive cut completely through the neck just below the head.

4. **Allow to bleed out.** The heart will continue pumping for a few moments (this is normal). Keep the bird in the cone until movements stop.

Using a Chopping Block:

1. **Lay the hen on her side** on a solid surface (like a tree stump or chopping block).
2. **Secure the body** by holding firmly or having a second person help.
3. **Position the neck.** Lay the neck flat and extended on the block.
4. **One swift blow.** Bring the hatchet or cleaver down firmly and decisively in one motion, severing the head completely.
5. **Hold the body** for a few seconds as reflexive movements occur.

Important Notes:

- The blade **must** be extremely sharp. A dull blade requires multiple blows and causes unnecessary suffering.
- **One clean cut** is critical—practice your aim beforehand if needed (using a vegetable or similar object).
- Post-death reflexes (twitching, wing flapping) are normal and do not indicate pain or consciousness.
- This method is graphic and not suitable for everyone. Be honest with yourself about your comfort level.

Pros:

- Instantaneous when done correctly
- Traditional method used on farms for generations
- No special equipment beyond a sharp blade

Cons:

- Graphic and emotionally intense
- Requires sharp tools and confidence
- Reflexive movements can be disturbing
- Risk of injury to yourself if not careful

Aftercare: What to Do Next

Once euthanasia is complete:

1. **Confirm death.** Check for breathing, heartbeat (place hand on chest), and corneal reflex (gently touch the eye—it should not blink).
2. **Handle the body respectfully.** Wrap in cloth or place in a box if desired.
3. **Dispose according to local regulations:** Burial (2-3 feet deep), cremation, composting, or rendering services.
4. **Clean tools and area.** Disinfect any equipment used and wash your hands thoroughly.
5. **Take time for yourself.** This is emotionally taxing. Allow yourself to grieve and process what you've done.

Aftercare Options:

Burial:

- Dig a hole at least 2-3 feet deep to prevent scavengers from disturbing the body
- Wrap the hen in cloth or place her in a small box
- Mark the spot with a stone, plant, or marker if you wish
- Choose a meaningful location (under a favorite tree, near the garden)

Cremation:

- Some pet cremation services accept chickens
- You can keep the ashes or scatter them in your garden
- Costs vary but are typically $50-$150

Composting:

- Chickens can be composted in a deep, hot compost pile (not recommended for beginners)

- This method returns nutrients to the earth and may appeal to homesteaders

Rendering or Disposal:

- Some areas offer livestock disposal services
- Check local regulations for proper disposal methods
- Your vet may offer disposal services if you choose veterinary euthanasia

A Final Word

Performing euthanasia at home is never easy, but sometimes it's necessary. If you've followed these instructions carefully and acted with compassion, you've done right by your hen. You ended her suffering quickly and humanely, and that is an act of profound kindness.

If you found this process too difficult or traumatic, there's no shame in choosing professional veterinary care in the future. Every keeper must find the approach that aligns with their abilities and values.

Your hen was lucky to have you.

Grab your FREE gifts!

Sophie McKay's Seed Starting & Planting Calculator + The Ultimate Guide to Organic Weed Management

In these free resources, you will discover:

- The perfect Seed Starting and Planting times for YOUR region or zone
- The 8 Organic Weed Removal Methods
- The 6 best and proven Weed Management Methods
- The tools you did NOT know you need for a weed-free garden
- How weeds can help your yard
- How to identify which weed is good and which is bad for a yard or garden
- The difference between Invasive and Noxious Weeds

Get your FREE copy today by visiting:

https://sophiemckay.com/free-resources/

Unlock the Secrets to Thriving Fruit Tree Gardens!

Transform your backyard orchard dreams into reality with 'Beginner's Guide to Growing Fruit Trees Fast and Easy.' Your guide on this road will be Sophie McKay, an avid gardener and an emerging author in gardening, permaculture, and sustainability. She'll share her best tips and tricks to ensure your gardening journey succeeds.

From **efficient** garden **layout design to selecting healthy trees, introducing pruning and grafting basics, mastering sustainable pest management, and creating a permaculture-inspired food forest**—this guide is your go-to resource for cultivating a vibrant and fruitful orchard. With practical insights, rejuvenating techniques, and seasonal care tips,

embark on a sustainable gardening success story.

Just **scan this QR code** with your phone, or visit the https://BuyFTG.SophieMckay.com link to land directly on the book's Amazon page.

If You Liked This Book, Try This One Too!

Sophie's fantastic new book, **The Beginner's Guide to Successful Container Gardening,** is now published!

Inside, you will learn about the basics of container gardening, including selecting the right container, soil, and plants for all your needs. You will also learn about the specific requirements of different types of plants, and how to care for them throughout the growing season. Whether you're a seasoned gardener or just getting started, this book has something for everyone.

So if you're ready for some more inspiration, check out this book now to keep your garden thriving all year round with 25+ proven DIY methods for composting, companion planting, seed saving, water management and pest control!

So what are you waiting for? Grab it for yourself!

Just **scan this QR code** with your phone, or visit the https://Container.SophieMckay.com link to land directly on the book's Amazon page.

The Happy Hen Handbook

Welcome to Permaculture!

Unlock the secrets of a resilient garden! Discover permaculture design and **learn how to grow your own food in harmony with nature**.

Join Sophie on a guided tour and create your own **sustainable permaculture garden** with confidence. Success guaranteed!

Just scan this QR code with your phone, or visit the https://book.SophieMckay.com link to land directly on the book's Amazon page.

Bibliography

Food & Wine. (2025, February 10). *Concerned about getting eggs? Here's how you can rent a chicken*. Food & Wine. Retrieved June 23, 2025, from https://www.foodandwine.com/egg-shortage-chicken-rentals-8788563

Harrington, C. (2025, March 17). *Why Americans care so much about egg prices—and how this issue got so political*. The Conversation. Retrieved June 23, 2025, from https://theconversation.com/why-americans-care-so-much-about-egg-prices-and-how-this-issue-got-so-political-251752

McCausland, C. (2022, September). Thanks to the pandemic, people are flocking to a new trend: Backyard chicken-raising. *Baltimore Magazine*. Retrieved June 23, 2025, from https://www.baltimoremagazine.com/section/homegarden/baltimoreans-embrace-backyard-chicken-raising-trend/

Gates, B. (2016, June 7). *Why I would raise chickens*. Gates Notes. Retrieved June 23, 2025, from https://www.gatesnotes.com/why-i-would-raise-chickens

General Code. (2022). *Backyard chickens legislation*. General Code. Retrieved June 24, 2025, from https://www.generalcode.com/blog/backyard-chickens-legislation/

Smith, G., & Dunipace, S. (2011). How backyard poultry flocks influence the effort required to curtail avian influenza epidemics in commercial poultry flocks. *Epidemics*, *3*(2), 71–75. https://doi.org/10.1016/j.epidem.2011.01.003

Jacob, J. (2015). Developing regulations for keeping urban chickens. *eXtension*. Retrieved June 24, 2025, from https://poultry.extension.org/articles/poultry-management/urban-poultry/developing-regulations-for-keeping-urban-chickens/

Loeffler, B. (2024, November 20). *How does light affect egg production?* Redmond Agriculture Blog. https://blog.redmondagriculture.com/how-does-light-affect-egg-production

Alabama Cooperative Extension System. (2025, May 29). *Nutrition for backyard chicken flocks*. https://www.aces.edu/blog/topics/farming/nutrition-for-backyard-chicken-flocks/

Fowler, J. (2025, August 13). *Nutrition for the backyard flock*. CAES Field Report. https://extension.uga.edu/publications/detail.html?number=C954&title=nutrition-for-the-backyard-flock

Annie. (2025, May 21). What you need to know about chicken grit. *Strong Animals*. https://www.getstronganimals.com/post/what-you-need-to-know-about-chicken-grit?srsltid=AfmBOorO53FE7L_Dta5E64YjNfnrzt4irXjl14axlvws1WYwhE-5tXpg

PoultryDVM. (n.d.). *Giving oregano to chickens: Research, dosage and benefits*. Retrieved October 19, 2025, from https://poultrydvm.com/supplement/oregano

PoultryDVM. (n.d.). *Giving garlic to chickens: Research, dosage and benefits*. Retrieved October 19, 2025, from https://poultrydvm.com/supplement/garlic

British Hen Welfare Trust. (2024, January 10). *Apple cider vinegar for chickens – a natural way to good hen health*. Retrieved October 19, 2025, from https://www.bhwt.org.uk/blog/health-welfare/apple-cider-vinegar-for-chickens-a-natural-way-good-hen-health/

Kruk, D., Hanzu, I., & De Keukeleire, P. (2021). Water dynamics in eggs by means of nuclear magnetic resonance relaxometry: A study of water molecules embedded in egg yolk and white of three species. *Food Chemistry, 356*, Article 129659. https://doi.org/10.1016/j.foodchem.2021.129659

Lesley, C. (2021, January 17). *The complete guide to chickens and water*. Chickens and More. Retrieved October 19, 2025, from https://www.chickensandmore.com/chickens-and-water

Myers, M., & Ruxton, C. H. S. (2023). Eggs: Healthy or risky? A review of evidence from high quality studies on hen's eggs. *Nutrients, 15*(12), 2657. https://doi.org/10.3390/nu15122657

Sato, Y. (2025, April 7). *Common infectious diseases in backyard poultry*. MSD Veterinary Manual. https://www.msdvetmanual.com/exotic-and-laboratory-animals/backyard-poultry/common-infectious-diseases-in-backyard-poultry

McKay, S. (2022). The Practical Permaculture Project.

McKay, S. (2023). Beginners Guide to Successful Container Gardening.

Purina Mills. (n.d.). *When do chickens start laying eggs?* Retrieved October 19, 2025, from https://www.purinamills.com/chicken-feed/education/detail/when-do-chickens-start-laying-eggs

Annie. (2025, March 3). *What is the Bloom on an Egg? Strong Animals.* https://www.getstronganimals.com/post/what-is-the-bloom-on-an-egg?srsltid=AfmBOopxqj9-C75v6Zc67qJVT00G3pwjejFOVMQHDdckZ7sM3Ag_gutM

How to do the egg float Test (And why it works). (2025, August 14). Australian Eggs. https://www.australianeggs.org.au/facts-and-tips/egg-float-test

Trust, B. H. W. (2025, September 15). Why do hens eat their eggs? | British Hen Welfare Trust. *British Hen Welfare Trust.* https://www.bhwt.org.uk/blog/health-welfare/why-do-hens-eat-their-eggs/

purinamills.com. (2025, July 25). *Signs of a Healthy, happy chicken| Purina Animal Nutrition.* https://www.purinamills.com/chicken-feed/education/detail/is-my-flock-healthy-6-signs-of-happy-backyard-chickens

Nutri-Vet. (2025, August 8). 14 Common chicken diseases, Symptoms, Prevention and treatment. *MannaPro.* https://mannapro.com/blogs/news/top-14-chicken-diseases

Jacob, J. (n.d.). *External parasites of poultry – Small and backyard poultry.* Retrieved October 19, 2025, from https://poultry.extension.org/articles/poultry-health/external-parasites-of-poultry

Jacob, J. (n.d.). *Internal parasites of poultry – Small and backyard poultry.* Retrieved October 19, 2025, from https://poultry.extension.org/articles/poultry-health/internal-parasites-of-poultry/

Sanderson, R. (2023, April 6). *Sick Chicks: 7 common illnesses you may encounter.* Backyard Poultry. https://backyardpoultry.iamcountryside.com/feed-health/sick-chicks-7-common-illnesses-you-may-encounter/

www.ingramcontent.com/pod-product-compliance
Lightning Source LLC
Chambersburg PA
CBHW020408080526
44584CB00014B/1221